Lecture Notes in Mathematics

Edited by A. Dold and B. Eckmann

728

Non-Commutative Harmonic Analysis

Proceedings, Marseille-Luminy, France, June 26 to 30, 1978

Actes du Colloque d'Analyse Harmonique Non Commutative

Edited by
Jacques Carmona and Michèle Vergne

Springer-Verlag
Berlin Heidelberg New York 1979

Editors

Jacques Carmona
Université d'Aix-Marseille II
U.E.R. Scientifique de Luminy
70, Route Léon Lachamp
F-13288 Marseille Cédex 2

Michèle Vergne
Université Paris VII
U.E.R. de Mathématiques
2, Place Jussieu
F-75221 Paris Cédex 05

AMS Subject Classifications (1970): 16A66, 17B10, 17B35, 20G20, 22E30, 22E45, 32M15, 35E10, 43A80, 43A90

ISBN 3-540-09516-0 Springer-Verlag Berlin Heidelberg New York
ISBN 0-387-09516-0 Springer-Verlag New York Heidelberg Berlin

Library of Congress Cataloging in Publication Data
Colloque d'analyse harmonique non commutative, 3d, Marseille, 1978.
Commutative harmonic analysis. (Lecture notes in mathematics ; 728) English or French.
Bibliography: p. Includes index. 1. Harmonic analysis--Congresses. 2. Lie algebras--Congresses. 3. Lie groups--Congresses. I. Carmona, Jacques, 1934- II. Vergne, Michèle. III. Title. IV. Series: Lecture notes in mathematics (Berlin); 728.
QA3.L28 no. 728 [QA403] 510'.8s [515'.2433]
ISBN 0-387-09516-0 79-17579

2141/3140-543210

PREFACE

La troisième rencontre d'Analyse Harmonique Non Commutative sur les Groupes de Lie a eu lieu à Marseille-Luminy, du 26 au 30 Juin 1978, dans le cadre des activités du Centre International de Rencontres Mathématiques.

Ce volume contient le texte des Conférences présentées durant le Colloque à l'exception de celles qui concernaient des travaux déjà publiés ou dont la publication était prévue par ailleurs.

Outre les participants à cette rencontre, nous tenons à remercier l'U. E. R. Scientifique de Marseille-Luminy et le Centre International de Rencontres Mathématiques qui ont rendu possible la tenue de ce Colloque.

Jacques CARMONA
Michèle VERGNE

TABLE DES MATIERES

* to appear elsewhere

FOURIER TRANSFORMS OF SOME INVARIANT DISTRIBUTION ON SEMISIMPLE LIE GROUPS AND LIE ALGEBRAS

Dan Barbasch*
Institute for Advanced Study
Princeton, NJ 08540

Let G be a connected semisimple Lie group and $u \in G$ a unipotent element. Then the orbit of u by the adjoint action of G defines an invariant tempered distribution which we denote by T_u. In this paper we propose to investigate the explicit relation between the distribution T_u and the measures associated to semisimple orbits, more precisely the distribution F_f defined by Harish-Chandra. These results make it possible to calculate the Fourier transform of T_u in the sense of Harish-Chandra, i.e. to expand T_u in terms of distributions coming from characters of irreducible unitary representations. As motivation for this problem we consider the following situation. Let $\Gamma \subseteq G$ be a discrete subgroup such that G/Γ is compact. Let π be the left regular representation of G on $L^2(G/\Gamma)$. Then the well known Selberg trace formula states that

$$(1) \qquad \operatorname{tr} \pi(f) = \sum_{\gamma \in C} \operatorname{vol}(G_\gamma/\Gamma_\gamma) T_\gamma(f)$$

where $f \in C_c^\infty(G)$ and C is the set of conjugacy classes of elements in Γ by the adjoint action of G. An important problem for harmonic analysis and number theory is to find the multiplicities of irreducible representations occuring in $L^2(G/\Gamma)$. This can be done by observing that

$$(2) \qquad \operatorname{tr} \pi(f) = \sum_{\omega \in \mathcal{E}(G)} m_\omega \operatorname{tr} \omega(f)$$

where $\mathcal{E}(G)$ is the set of equivalence classes of irreducible unitary representations and m_ω the multiplicities. If we can find the Fourier transform in the sense of Harish-Chandra of each T_γ i.e. find an explicit formula

$$(3) \qquad T_\gamma(f) = \int_{\mathcal{E}(G)} \operatorname{tr} \omega(f) d\mu(\omega)$$

then we can calculate the m_ω explicitly.

Due to the work of Arthur [1], Warner [2] and Warner and Osborne [4] on the Selberg trace formula for $\operatorname{vol}(G/\Gamma) < \infty$, distributions of the form T_γ with γ unipotent occur. Formulas of type (3) were obtained for γ semisimple by Sally & Warner [3], Herb [5] and Chao [6]. The corresponding formulas for γ

---

*Supported in part by a National Science Foundation Grant.

unipotent are simple consequences of these formulas and the relations in theorems 2 and 3.

The measures T_Y have obvious analogues for the Lie algebra. Let B be the Cartan-Killing form. Then we can consider the Euclidean Fourier transform $\hat{T}_X$ for any $X \in \mathfrak{g}$, with respect to B. Then $\hat{T}_X$ is an invariant eigendistribution for G-invariant constant coefficient differential operators. By using the results of Rossmann [9] it is then possible to obtain formulas for $\hat{T}_X$ when X is nilpoter

These questions are completely answered in the case real rank $\mathfrak{g} = 1$ in Barbasch [8]. The generalizations proved here are inspired by the results in that paper.

Before we state the main results we prove a general lemma.

<u>Lemma 1</u>: Let $\mathfrak{p} = \mathfrak{m} + \mathfrak{a} + \mathfrak{n}$ be a parabolic subalgebra with its corresponding Langlands decomposition and P = MAN the corresponding parabolic subgroup. Assume that there is $X \in \mathfrak{n}$ such that $P \cdot X$ is open and dense in $\mathfrak{n}$. Let $H \in \mathfrak{a}$ be such that $\mathrm{Cent}_{\mathfrak{g}} H = \mathfrak{m} + \mathfrak{a}$ and $r = \dim \mathrm{Cent}_{\mathfrak{g}} H$. Then

$$(4) \qquad \lim_{t \to 0^+} t^{(n-r)/2} T_{tH}(f) = \pi(H)^{-1} c_X T_X(f)$$

where $\pi(H) = |\det_n(\mathrm{ad}\, H)|$ and c_X depends only on the normalization of the measures.

<u>Proof</u>: This is well known. We sketch a proof for completeness. Let $\mathfrak{g} = \underline{k} + \mathfrak{s}$ be a Cartan decomposition and K the group corresponding to $\underline{k}$. Then we may assume that $f(k \cdot Z) = f(Z)$ for any $k \in K$. Since $G = K \cdot P$ we can write

$$(5) \qquad T_X(f) = \int_P f(p \cdot X) d\mu(p) = \int_n f(Z) dZ$$

and

$$(6) \qquad T_{tH}(f) = \int_N f(tn \cdot H) = \pi(tH)^{-1} \int_n f(tH + Z) dZ \quad .$$

Since $\dim n = \frac{n-r}{2}$, the proof is complete.

As a consequence we have the following theorem.

<u>Theorem 2</u>: Let $\mathfrak{g} = sl(n, \mathbb{R})$ and $\{X, H, Y\}$ be a Lie triple.

a) If ad H has odd eigenvalues then there is a parabolic $\mathfrak{p}_X$ such that X satisfies the conditions in Lemma 1.

b) If ad H has even eigenvalues only, then

$$(7) \qquad \lim_{t \to 0^+} t^{(n-r)/2} T_{t(X-Y)}(f) = c_X T_X(f) \quad .$$

<u>Proof</u>: In view of Lemma 1 we have to determine which $X \in \mathfrak{g}$ nilpotent satisfy the condition in Lemma 1. Of course part b) is more difficult because X-Y is an elliptic element. The existence of such a formula is motivated by the

fact that the conjugacy class of X (or equivalently $\{X, H, Y\}$) with respect to G is determined by the G-conjugacy class of $X-Y$. (A proof of this can be found in Barbasch [ ]. It was first observed by R. Rao and then proved by B. Kostant.)

We first consider conjugacy classes of nilpotents with respect to $GL(n)$. Representatives are given by the Jordan canonical form. This consists of boxes of the type

$$n_j \overbrace{\begin{bmatrix} 0 & 1 & & \\ & \cdot & \cdot & \\ & & \cdot & \cdot \\ & & & \cdot\ \cdot\ 1 \\ 0 & & & 0 \end{bmatrix}}^{n_j}$$

(where n_j could be one). Then the conjugacy classes of nilpotents are parametrized by partitions

$$n_1 \geq \cdots \geq n_s \geq 0, \qquad n_1 + \cdots + n_s = n \quad .$$

On the other hand let $\mathfrak{P}$ be a parabolic subalgebra such that $m + \mathfrak{a} = gl(m_1) \times \cdots \times gl(m_k)$ where $m_1 + \cdots + m_k = n$ and $m_k \leq \cdots \leq m_1$. Then it is straightforward to see that the nilpotent with the open orbit in n is given by m_k boxes of size k, $(m_{k-1} - m_k)$ boxes of size $k-1$, etc. Since this exhausts all possible conjugacy classes of nilpotents, we can apply Lemma 1 in this case. As a consequence we also get the fact that any nilpotent $X \in sl(n, \mathbb{R})$ has an open orbit in some n for a parabolic.

The conjugacy classes of nilpotents with respect to $SL(n)$ are the same as for $GL(n)$ except for the case when all the n_j's are even. In that case X and $-X$ are not conjugate with respect to $SL(n)$. In this case, since H is formed of diagonal boxes

$$\begin{bmatrix} n_j+1 & & & \\ & n_j-1 & & \\ & & \cdot & \\ & & & \cdot \\ & & & & -n_j-1 \end{bmatrix}$$

and it becomes clear that $\operatorname{ad} H$ has only even eigenvalues. (This is relevant for example for the fact that $\dim \operatorname{Cent} H = \dim \operatorname{Cent} X = \dim \operatorname{Cent} (X-Y)$ as can be seen from a calculation of the representations of $sl(2) \approx \{X, H, Y\}$ occuring in $\mathfrak{g}$.)

We now recall that Harish-Chandra has done a very deep analysis of the distribution

$$F_f^L(b) = \Delta_L(b) \int_{G/L_R} f(x \cdot b \cdot x^{-1})dx = \Delta_L(b) T_b(f) \quad .$$

Here L is a Cartan subgroup, $b \in L'$ an element in the regular set and $\Delta_L(b)$ a certain determinant function. A convenient reference for basic results about F_f, for the group as well as their analog for $\mathfrak{g}$, can be found in Varadarajan [ ]. In particular we need the fact that, for any Z semisimple, $T_{tZ}(f)$ has an asymptotic expansion in t. In particular the first nonzero term in this expansion is a measure. We have to show that T_X is the only distribution occuring in the first nonzero term. We note that, if X corresponds to the partition $n_1 \geq n_2 \geq \ldots \geq n_k$ then $X-Y$ can be conjugated into an element of the type

$$\pm \begin{bmatrix} & \cdot^{\cdot^{\cdot}} n_1+1 & & & \\ -(n_1+1) & \vdots & & & \\ \cdot^{\cdot^{\cdot}} & \vdots & & & \\ \cdot \;\cdot\;\cdot & \cdot & \cdot\;\cdot\;\cdot & \cdot^{\cdot^{\cdot}} n_1-1 & \\ & \vdots & -(n_1-1) \cdot^{\cdot^{\cdot}} & \vdots & \\ & \cdot & \cdot\;\cdot\;\cdot & \cdot & \text{etc.} \end{bmatrix}$$

Since the centralizer of this element is contained in the Levi component of a parabolic defined by the size of the boxes involved, we can write down the measure corresponding to the orbit of $t(X-Y)$ explicitly (in terms of the decomposition $G = K \cdot P$) and then calculate the limit.

This completely solves the problem of calculating Fourier transforms of nilpotent orbits in the case $\mathfrak{g} = sl(n, \mathbb{R})$.

The case of a general semisimple Lie algebra is much more complicated. In the previous formulas the semisimple orbit had the same dimension as the nilpotent orbit. As can be seen from Barbasch [8] this cannot always be done, even when real rank $\mathfrak{g} = 1$.

The next theorem, treating some special nilpotents in $\mathfrak{g} = sp(n)$ illustrates what the difficulties are.

<u>Theorem 3</u>: Let $\mathfrak{g} = sp(n, \mathbb{R})$.

a) Let $X_{p,q} = \begin{bmatrix} 0 & A \\ 0 & 0 \end{bmatrix}$ and $Z_{p,q} = \begin{bmatrix} 0 & A \\ -A & 0 \end{bmatrix}$ where $A = \begin{bmatrix} I_p & 0 \\ 0 & -I_q \end{bmatrix}$ and $p + q = n$. Then

(8) $$\lim_{t \to 0^+} t^{n(n+1)/2} \, T_{tZ_{p,q}}(f) = T_{X_{p,q}}(f) .$$

b) Let $X_{\alpha,\beta,0} = \begin{bmatrix} 0 & A \\ 0 & 0 \end{bmatrix}$ where $A = \begin{bmatrix} I_\alpha & & \\ & -I_\beta & \\ & & 0 \end{bmatrix}$, $\alpha + \beta = n - 1$. Then

(9) $$\frac{1}{2} \frac{d}{dt}\Big|_{t=0^+} t^{n(n+1)/2} \sum_{\gamma=n-\beta}^{n-1} (-1)^{\gamma-\beta} T_{tZ_{\gamma+1,n-\gamma-1}}(f) = T_{X_{\alpha,\beta,0}}(f) .$$

Proof: The proof of these formulas follows by finding a transverse to the semisimple orbit and expressing the measure with respect to that transverse. The simplest case is when A is positive definite. Then the KAK decomposition is sufficient. Rather than go through the proof we give a different one for $\mathfrak{g} = sp(2, \mathbb{R})$ which illustrates the role played by the geometry of the orbits. In this case the conjugacy classes of nilpotents are given by

$$1)\quad X = \pm\begin{bmatrix} 0 & \sqrt{3} & 0 & 0 \\ 0 & 0 & 0 & 2 \\ 0 & 0 & 0 & 0 \\ 0 & 0 & -\sqrt{3} & 0 \end{bmatrix}, \quad H = \begin{bmatrix} 3 & 0 & 0 & 0 \\ 0 & 1 & 0 & 0 \\ 0 & 0 & -1 & 0 \\ 0 & 0 & 0 & -3 \end{bmatrix}, \quad Y = \pm\begin{bmatrix} 0 & 0 & 0 & 0 \\ \sqrt{3} & 0 & 0 & 0 \\ 0 & 0 & 0 & -\sqrt{3} \\ 0 & 2 & 0 & 0 \end{bmatrix}.$$

$$2)\quad X = \begin{bmatrix} 0 & A \\ 0 & 0 \end{bmatrix}, \quad H = \begin{bmatrix} I & 0 \\ 0 & -I \end{bmatrix}, \quad Y = \begin{bmatrix} 0 & 0 \\ A & 0 \end{bmatrix}$$

where $A = +I$, $A = -I$ or $A = [\begin{smallmatrix} 1 & 0 \\ 0 & -1 \end{smallmatrix}]$.

$$3)\quad X = \begin{bmatrix} 0 & A \\ 0 & 0 \end{bmatrix}, \quad H = \begin{bmatrix} 1 & & & \\ & 0 & & \\ & & -1 & \\ & & & 0 \end{bmatrix}, \quad Y = \begin{bmatrix} 0 & 0 \\ A & 0 \end{bmatrix}$$

where $A = \pm[\begin{smallmatrix} 1 & 0 \\ 0 & 0 \end{smallmatrix}]$.

The cases listed in 2 and 3 with + and - are not conjugate. We consider case 3 with +.

Let $z_y = \text{Cent}_{\mathfrak{g}} Y$. Then $\underline{U} = x + z_y$ is, according to some unpublished results of R. Rao, a transverse to the adjoint action of G in particular to any orbit intersecting it. Let $\beta \in C_c^\infty(G \times \underline{U})$. Then there is a well defined correspondence (integration on fibers) $\beta \longmapsto f_\beta$ where $f_\beta \in C_c^\infty(G\cdot\underline{U})$ which is onto. Then for $Z \in \underline{U}$,

$$(9)\qquad \int_{G/G_Z} f_\beta(xZx^{-1})dx = \int_G \int_{\underline{U}\cap O(Z)} \beta(x, u)d\omega(u)dx .$$

Let $r = \dim \text{Cent}_{\mathfrak{g}} X = \dim \underline{U}$. Then

$$(10)\qquad f_\beta(tv) = t^{-(n-r)/2} f_{\beta_t}(v)$$

where $\beta_t = \beta\circ\phi$ and $\phi(x, u) = (x\gamma_t, u_t)$ and $\gamma_t = \exp(-\frac{1}{2}\log tH)$ and $u_t = X + t\gamma_t\cdot u_y$.

We now calculate the intersection $O(X - Y) \cap \underline{U}$. $X - Y$ satisfies the equation $x(x^2 + 1) = 0$. We get two orbits satisfying this equation:

$$\begin{bmatrix} 0 & 0 & 1 & 0 \\ 0 & 0 & 0 & 0 \\ -1 & d & 0 & 0 \\ d & -d^2 & 0 & 0 \end{bmatrix}, \quad d \in \mathbb{R} \quad \text{and} \quad \begin{bmatrix} 0 & 0 & 1 & 0 \\ 0 & a & 0 & b \\ 0 & 0 & 0 & 0 \\ 0 & c & 0 & -a \end{bmatrix}, \quad a^2 + bc + 1 = 0 .$$

The first one is $O(X - Y) \cap \underline{U}$, the second is $O(Z_0) \cap \underline{U}$ where $Z_0 = \left[\begin{smallmatrix} 0 & I \\ -I & 0 \end{smallmatrix}\right]$.

The action of the centralizer of $\{X, H, Y\}$ ($\approx sp(1)$) leaves $\underline{U}$ invariant. Then identifying $sp(1)$ with $sl(2, \mathbb{R})$, the first orbit is the orbit of

$$X_0 = \begin{bmatrix} 0 & 0 & 1 & 0 \\ 0 & 0 & 0 & 0 \\ -1 & 1 & 0 & 0 \\ 1 & -1 & 0 & 0 \end{bmatrix}$$

which has centralizer the same as $\left[\begin{smallmatrix} 0 & 1 \\ 0 & 0 \end{smallmatrix}\right]$ in $sl(2, \mathbb{R})$ and the second orbit corresponds to

$$Z_0' = \begin{bmatrix} 0 & 0 & 1 & 0 \\ 0 & 0 & 0 & 1 \\ 0 & 0 & 0 & 0 \\ 0 & -1 & 0 & 0 \end{bmatrix}$$

which corresponds to $\left[\begin{smallmatrix} 0 & 1 \\ -1 & 0 \end{smallmatrix}\right]$ in $sl(2, \mathbb{R})$.

Since $O(X) \cap \underline{U} = \{X\}$, the measure corresponding to $O(X)$ on this transverse is $dx \otimes \delta_X$ where dx is the Haar measure and δ_X the delta function at X.

Thus by using formulas (8) and (9) we get

$$t^3 \int_{G/G_{X-Y}} f_\beta(tx\cdot(X-Y))dx = t \int_G \int_{SL(2)/G_{X_0}} \beta(x, X-t^2Y \pm t\cdot g\cdot X_0)dgdx .$$

Since X_0 is nilpotent we can never get $\int_G \beta(x, X)dx$ in the limit no matter what differential operator we use.

On the other hand, for Z_0 we get

$$t^3 \int_{G/G_{Z_0}} f_\beta(tx\cdot Z_0)dx = t \int_G \int_{SL(2)/G_{Z_0'}} \beta(x, X-t^2Y + tg\cdot Z_0')dgdx .$$

But then the formula stated in the theorem reduces to the well known statement that

$$F_f^B(0; \omega) = cf(0), \qquad c \neq 0$$

for the Lie algebra $sl(2, \mathbb{R})$. In order to have a complete proof we need to do some more work since $G\cdot\underline{U}$ is open in $\mathfrak{g}$ but not equal to it. We omit the details.

Finally, we remark that certain invariant distributions with support in the cone of nilpotent elements are associated to irreducible representations. The relationship involves the euclidean transforms of semisimple and nilpotent orbits and limits measures that are considered in this paper. We refer to Kashiwara &

Vergue 10] and Barbasch & Vogan [11] for an exposition of these ideas.

References

(1) Arthur, J., The Selberg trace formula for groups of F rank one, Ann. of Math., vol. 100, 1974.

(2) Warner, G., The Selberg trace formula for Lie groups of real rank one, pre-print, to appear, Advances of Math.

(3) Sally, P. and Warner, G., The Fourier transform on semisimple Lie groups of real rank one, Acta Math., vol. 131, 1973.

(4) Osborne, S. and Warner, G., Multiplicities of integrable discrete series, The case of a nonuniform lattice in an R-rank one group, to appear.

(5) Herb, B., Fourier inversion of invariant integrals on semisimple real Lie groups, preprint.

(6) Chao, W., Fourier inversion and the Plancherel formula for semisimple Lie groups of real rank two, Ph.D. Thesis, U. of Chicago, 1977.

(7) Varadarajan, V., Harmonic Analysis on real reductive groups, Lecture Notes in Math. 576, Springer Verlag.

(8) Barbasch, D., Fourier inversion for unipotent invariant integrals, to appear, Transactions of AMS.

(9) Rossmann, W., Kirillov's Character formula for semisimple Lie groups, preprints.

(10) Kashiwara, M. and Vergne, M., K-types and singular spectrum, same volume.

(11) Barbasch, D. and Vogan, D., The local structure of characters, preprint.

GLOBAL SOLVABILITY OF BI-INVARIANT DIFFERENTIAL

OPERATORS ON SOLVABLE LIE GROUPS

Weita Chang

§1. Main result

The purpose of this note is to announce the following C^∞-global solvability result. (Differential operators are always linear in this note.)

Theorem. Let G be a simply connected solvable Lie group, P a non-zero bi-invariant differential operator on G. Then for any C^∞ function f on G, we can find a C^∞ function u on G such that $Pu = f$ holds on G. (This property will simply be called the global solvability of P).

As is well known in the general theory of P.D.E., the conjunction of semi-global solvability (solvability on each compact set) of P and P-convexity (see §3 for the definition) will imply the above mentioned global solvability. In [10], Rouvière proved semi-global solvability of bi-invariant differential operators on simply connected solvable Lie groups. So all we have to show is P-convexity. In §4, we will describe the main steps in our proof of P-convexity. The complete details will appear in [3]. In §2, §3, we will make historical comments and give some explanation of P-convexity.

§2. Historical comments

Let us review here some of known results on solvability of invariant differential operators.

Differential operators with constant coefficients are globally solvable on $\mathbb{R}^n$. In fact, for any open convex set $\Omega \subset \mathbb{R}^n$, and a non-zero constant coefficient operator P, we have $PC^{\infty}(\Omega) = C^{\infty}(\Omega)$ (c.f. [8].) On a general Lie group, left-invariant differential operators are not always even locally solvable (Cerèzo-Rouvière [2]). In [2], one can find a necessary and sufficient condition on a group for all its left-invariant differential operators to be locally solvable. But for bi-invariant operators, Duflo [5] proved local solvability on arbitrary Lie groups. Turning to global solvability of bi-invariant differential operators, again the answer is negative in general. Rouvière pointed out that every complex semi-simple Lie group has a bi-invariant operator which is not globally solvable ("the imaginary Casimir"). On the other hand, we know that the Casimir operator on a non-compact semi-simple Lie group with finite center is globally solvable. This is a result by Rauch-Wigner [9]. Recently Chang [4] has extended their result to global solvability of the Laplacian on a pseudo-Riemannian symmetric space G/H where G is a non-compact semi-simple Lie group with finite center and H is an open subgroup of the fixed point group of an involutive automorphism of G. If H is maximal compact so that G/H is a Riemannian symmetric space of non-compact type, Helgason [7] proved global solvability of G-invariant operators on G/H. In [11], Wigner showed global solvability of bi-invariant differential operators on simply connected nilpotent Lie groups by demonstrating P-convexity. Our proof of P-convexity in the solvable case uses a generalization of his method. We remark that Duflo-Wigner [6] independently proved P-convexity in the solvable case. Also they obtained P-convexity results for more general groups.

§3. Notion of P-convexity

In this section we explain the significance of P-convexity. Let M be a smooth manifold and P a differential operator with smooth coefficients on

M. $C_0^\infty(M)$ shall denote the space of smooth functions with compact support on M.

<u>Definition.</u> M is called P-convex if for each compact set $K \subset M$, there exists a compact set $K' \subset M$ such that for each distribution u with compact support on M, $\text{supp}\, {}^tPu \subset K$ implies $\text{supp}\, u \subset K'$. Here ${}^tP$ is the transpose of P with respect to a nowhere vanishing smooth measure on M and "supp" means support.

It is well known that if the manifold M is countable at infinity, the conjunction of P-convexity of M and semi-global solvability (i.e., $PC^\infty(M) \supset C^\infty(K)$ for each compact K) gives global solvability.

Let us give a simple example. Let M be the open set in $\mathbb{R}^2$ surrounded by the curve as shown in Figure 1.

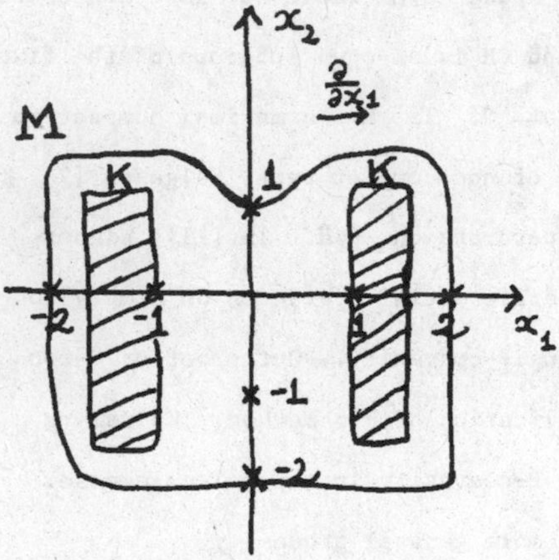

Figure 1

Observe that M is not convex in the x_1-direction. We are going to show that M is not $\frac{\partial}{\partial x_1}$-convex. Choose $\varphi, \rho_n \in C_0^\infty(\mathbb{R})$ with the following properties.

$$\varphi(t) \equiv 1 \qquad -1 \leq t \leq 1$$
$$\equiv 0 \qquad |t| \geq 1.5$$
$$\rho_n(t) \equiv 1 \qquad -1 \leq t \leq 1 - 1/n$$
$$\equiv 0 \qquad t \leq -1.5,\ t \geq 1 - 1/2n$$

Put $u_n(x_1,x_2) = \varphi(x_1)\rho_n(x_2)$. Then $u_n \in C_0^\infty(M)$ for each integer n and $\frac{\partial}{\partial x_1}u_n = 0$ for $-1 \leq x_1 \leq 1$. It is easy to verify that there exists a compact set $K \subset M$ as shown in Figure 1 such that $\text{supp}\frac{\partial}{\partial x_1}u_n \subset K$ for all integer n.

Nevertheless there is no compact set in M which contains all the $\operatorname{supp} u_n$ because $u_n = 1$ at $(0,1 - 1/n)$ for each n. Thus M is not $\frac{\partial}{\partial x_1}$-convex. Moreover, we can see that $\frac{\partial}{\partial x_1}$ is not globally solvable on M. In fact, let $u \in C^\infty(M)$ and put $\frac{\partial}{\partial x_1}u = f$. Then $u(1,x_2) - u(-1,x_2) = \int_{-1}^{1} f(x_1,x_2)dx_1$ for $0 \leq x_2 < 1$. If f is positive and blows up at $(0,1)$, then the above integral will blow up as x_2 approaches 1. But $u(1,x_2) - u(-1,x_2)$ can not blow up as $x_2 \to 1$. Therefore if $\frac{\partial}{\partial x_1}u = f$ for some $u \in C^\infty(M)$, then f must be a function on M which does not blow up at $(0,1)$. Hence $\frac{\partial}{\partial x_1}$ is not globally solvable on M.

In order to prepare for §4 where a sketch of the proof of P-convexity of our main theorem is given, we recall the definition of the principal symbol of a differential operator and the uniqueness theorem of Holmgren.

<u>Definition</u>. Let M be a smooth manifold, T^*M its cotangent bundle. Let D be a differential operator of order m on M. The principal symbol $\sigma(D)$ of D is the map $T^*M \to \mathbb{C}$ given by

$$\sigma(D)(df(x)) = \frac{1}{m!}D_y(f(y) - f(x))^m\Big|_{y=x}$$

for real valued functions $f \in C^\infty(M)$, $x \in M$. Here df is the differential of f and D_y is the operator D acting on the y-variable.

<u>Remark</u>. Let G be a Lie group with Lie algebra $\mathfrak{g}$. We regard $\mathfrak{g}$ as the space of real left-invariant vector fields on G. Then every left-invariant differential operator P on G is written as

$$P = \sum_{|\alpha|\leq m} a_\alpha X_1^{\alpha_1} \cdots X_n^{\alpha_n} \quad , \quad a_\alpha \in \mathbb{C}$$

where m is the degree of P and $\alpha = (\alpha_1,\ldots,\alpha_n)$ $|\alpha| = \alpha_1 + \cdots + \alpha_n$. Then the principal symbol of P is given by $\sigma(P)(df(x)) = \sum_{|\alpha|=m} a_\alpha (X_1 f(x))^{\alpha_1} \dots (X_n f(x))^{\alpha_n}$.

The uniqueness theorem of Holmgren (c.f. [8], Theorem 5.3.1.) Let M be a real analytic manifold, D a differential operator with analytic coefficients on M. Let $x_0 \in M$ and $\varphi \in C^\infty(M)$ be a real valued function. If $\sigma(D)(d\varphi(x_0)) \neq 0$, then any distribution u on M satisfying $Du \equiv 0$ near x_0 and vanishing identically on $\{x \in \Omega | \varphi(x) > \varphi(x_0)\}$ for some open neighborhood Ω of x_0 must vanish identically near x_0.

§4. Proof of P-convexity

We will sketch the proof of P-convexity of a simply connected solvable Lie group for a bi-invariant operator P.

Throughout this section let G be a simply connected solvable Lie group with Lie algebra $\mathfrak{g}$. $U(\mathfrak{g})$ shall denote the enveloping algebra of $\mathfrak{g}$ which we identify with the algebra of left-invariant differential operators on G. Then its center $Z(\mathfrak{g})$ is the algebra of bi-invariant differential operators. We fix a right-invariant measure on G so that the transpose map $P \to {}^tP$ gives an anti-automorphism on $Z(\mathfrak{g})$. We call a closed set $F \subset G$ P-full for a differential operator P on G if for any distribution u on G with compact support, supp $Pu \subset F$ implies supp $u \subset F$. Now, it is easily seen that in order to prove our desired P-convexity it suffices to prove the following.

(4.1) Let P be a non-zero bi-invariant differential operator on G. Then every compact set $K \subset G$ is contained in a P-full compact set of G.

The proof of (4.1) is reduced to the following two lemmas.

<u>Lemma 1</u> (Wigner [11]). <u>Let</u> Z <u>be a connected closed central one-parameter subgroup of</u> G. <u>(Assume such a</u> Z <u>exists.) Let</u> $\tilde{P}$ <u>be a non-zero bi-invariant differential operator on</u> G, P <u>be the bi-invariant differential operator on</u> G/Z <u>given by restricting</u> P <u>to the right</u> Z-<u>invariant functions. Let</u> $K \subset G/Z$ <u>be compact and</u> $\tilde{P}$-<u>full. Then</u> $\pi^{-1}(K)$ <u>is</u> P-<u>full where</u> $\pi: G \to G/Z$ <u>is the canonical projection</u>.

<u>Lemma 2</u> (Chang [3]). <u>With the same notation as above, let</u> F <u>be a compact set of</u> G/Z. <u>Then there exists a real valued function</u> $\varphi \in C^{\infty}(G)$ <u>such that</u>

(i) $\sigma(P)(d\varphi) \neq 0$ on $\pi^{-1}(F)$

(ii) $X_n\varphi \equiv 1$ on G.

Here X_n is a generator of Z so that $\{\exp tX_n | t \in \mathbb{R}\} = Z$.

We now want to show how to deduce (4.1) from these two lemmas. We use induction on $\dim G$. Thus assume that (4.1) is true for all non-zero bi-invariant operators on simply connected solvable Lie groups of lower dimension than $\dim G$. (In case $\dim G = 1$, (4.1) is easy to verify.) Now let P be a non-zero bi-invariant differential operator on G. There are two cases to consider.

<u>Case I</u>. Assume that the center of $\mathfrak{g}$, the Lie algebra of G, is zero. In this case by a lemma of Borho ([1], page 58), there exists an ideal $\mathfrak{h}$ of codimension one in $\mathfrak{g}$ such that $P \in U(\mathfrak{h}) \subset U(\mathfrak{g})$. If H denotes the analytic subgroup of G corresponding to $\mathfrak{h}$, then we can regard P as an operator on H. Let K be a compact set of G. Then for some compact set $K_0 \subset H$, and a constant M, we can write

$$K \subset \{\exp tX_1 \cdot K_0 \,|\, |t| \leq M\}$$

where $X_1 \notin \mathfrak{h}$ so that the map $(t,h) \mapsto \exp tX_1 \cdot h$ gives a diffeomorphism of $\mathbb{R} \times H$ onto G. Applying the induction hypothesis to H and P, we find a P-full compact set K_1 in H containing K_0. Now the set $\{\exp tX_1 \cdot K_1 \,|\, |t| \leq M\}$ obviously contains K and is compact. Using the P-fullness of K_1, we can prove that it is actually P-full in G. Thus Case I is settled.

Case II. Assume that there exists a non-zero central element X_n in $\mathfrak{g}$. Put $Z = \{\exp tX_n \,|\, t \in \mathbb{R}\}$. Let ℓ be the integer (possibly zero) such that $P = P_1 \cdot X_n^{\ell}$ and $\tilde{P}_1 \neq 0$. (Recall that $\tilde{P}_1$ is the operator on G/Z induced from P). Let $K \subset G$ be a compact set. Its projection $\pi(K)$ to G/Z is compact. Applying our induction hypothesis to G/Z and $\tilde{P}_1$, there exists a compact $\tilde{P}_1$-full set F containing $\pi(K)$. By Lemma 1, $\pi^{-1}(F) \supset K$ is P_1-full. By Lemma 2, there exists a real valued function $\varphi \in C^{\infty}(G)$ such that $\sigma(P_1)(d\varphi) \neq 0$ on $\pi^{-1}(F)$ and $X_n\varphi \equiv 1$ on G. Set $E = \{x \in \pi^{-1}(F) \,|\, |\varphi(x)| \leq M\}$ where we choose M large so that $E \supset K$. E is compact because $\pi^{-1}(F)$ is unbounded only in the Z-direction and the condition $X_n\varphi \equiv 1$ implies that any set of the form $\{|\varphi(x)| < c\}$ is bounded in the Z-direction. (Here, our explanation is a little bit rough but we do not try to make everything precise right now.) We claim that E is P_1-full. Assume the contrary. Then there is a distribution u of compact support such that

$$\text{(4.2)} \qquad \operatorname{supp} P_1 u \subset E, \quad \operatorname{supp} u \not\subset E.$$

We can find $x_0 \in \operatorname{supp} u$ such that $|\varphi(x_0)| = \sup_{x \in \operatorname{supp} u} |\varphi(x)| > M$. Note that since $\pi^{-1}(F)$ is P_1-full, $\operatorname{supp} u \subset \pi^{-1}(F)$. Without loss of generality, we may assume that $\varphi(x_0) = \sup_{x \in \operatorname{supp} u} |\varphi(x)|$ so that $u \equiv 0$ on $\{x \in G \,|\, \varphi(x) > \varphi(x_0)\}$. Also since $\varphi(x_0) > M$ we have that $P_1 u \equiv 0$ near x_0. Recall that $\sigma(P_1)(d\varphi(x_0)) \neq 0$. The uniqueness theorem of Holmgren now implies that $u \equiv 0$

near x_0. This is a contradiction because $x_0 \in \operatorname{supp} u$. Therefore (4.2) can never happen. Thus E is P_1-full. On the other hand, E is convex in the Z-direction and from this we can deduce that E is X_n-full. Hence E is $P = P_1 \cdot X_n^{\ell}$-full. Thus Case II is settled. Now (4.1) is proved.

Finally, we remark that a generalization of the above technique yields the following result.

Let G be a completely solvable simply connected Lie group, P a non-zero left-invariant differential operator on G. Then G is P-convex.

Recall that G is completely solvable if its Lie algebra $\mathfrak{g}$ has a chain of ideals $\mathfrak{g} = \mathfrak{g}_0 \supsetneq \cdots \supsetneq \mathfrak{g}_n = \{0\}$ with $\dim \mathfrak{g}_{i-1}/\mathfrak{g}_i = 1$ for all i. The proof of the above result will appear in [3].

References

[1] W. Borho, P. Gabriel, R. Rentschler, Primideale in Einhüllenden auflösbarer Lie-Algebren, Lecture Notes in Math., 357, New York-Heidelberg-Berlin, Springer-Verlag, 1973.

[2] A. Cerèzo, F. Rouvière, Résolubilité locale d'un opérateur différentiel invariant du premier ordre, Ann. Sci. Ecole Norm. Sup., 4 (1971), 21-30.

[3] W. Chang, Invariant differential operators and P-convexity of solvable Lie groups (To appear).

[4] W. Chang, Global solvability of the Laplacians on pseudo-Riemannian symmetric spaces (To appear in Journal of Functional Analysis).

[5] M. Duflo, Opérateurs différentiels bi-invariants sur un groupe de Lie. Ann. Sci. Ecole Norm. Sup., 10 (1977), 265-288.

[6] M. Duflo, D. Wigner, Convexité pour les opérateurs différentiels invariants sur les groupes de Lie (preprint).

[7] S. Helgason, Surjectivity of invariant differential operators on symmetric spaces I, Ann. of Math., 98 (1973), 451-479.

[8] L. Hörmander, Linear partial differential operators, New York, Springer-Verlag, 1963.

[9] J. Rauch, D. Wigner, Global solvability of the Casimir operators, Ann. of Math. 103 (1976), 229-236.

[10] F. Rouvière, Sur la résolubilité locale des opérateurs bi-invariants. Annali Scuola Normale Superiore Pisa 3 (1976), 231-244.

[11] D. Wigner, Bi-invariant operators on nilpotent Lie groups, Inventiones Math. 41 (1977), 259-264.

Department of Mathematics
Yale University
New Haven, Connecticut 06520
USA

L. CLOZEL

"Base change" géométrique :

Relèvement de la série principale de $GL(n,\mathbb{C}/\mathbb{R})$.

Si F/E est une extension de corps locaux, la fonctorialité en théorie de Langlands implique en particulier une correspondance entre représentations de GL (n,E) et certaines représentations de GL(n,F).

On montre ici que pour la série principale de $GL(n,\mathbb{C})$ et $GL(n,\mathbb{R})$, cette correspondance passe bien par le "relèvement local" de Saito et Shintani. La démonstration utilise le "théorème de Lefschetz" d'Atiyah et Bott.

1 - Introduction

Soit $G_{\mathbb{R}} = GL(n,\mathbb{R})$, $G_{\mathbb{C}} = GL(n,\mathbb{C})$. Soit σ la conjugaison de $G_{\mathbb{C}}$ par rapport à $G_{\mathbb{R}}$.

Soit $W_{\mathbb{C}} = \mathbb{C}^{\times}$, $W_{\mathbb{R}} = \mathbb{C}^{\times} \ltimes \{\pm 1\}$ (produit semi-direct par l'action de $\{\pm 1\}$ sur $\mathbb{C}^{\times}$ comme groupe de Galois). $W_{\mathbb{R}}$ s'identifie à l'ensemble des matrices monomiales de la forme $\begin{pmatrix} a & \\ & \bar{a} \end{pmatrix}$ ou $\begin{pmatrix} & b \\ -\bar{b} & \end{pmatrix}$ $(a,b \in \mathbb{C}^{\times})$ dans la réalisation matricielle

$$\mathbb{H} = \left\{\begin{pmatrix} a & b \\ -\bar{b} & \bar{a} \end{pmatrix}\right\} \text{ du corps des quaternions.}$$

La dualité de Langlands dans le cas archimédien ("classification de Langlands") donne une correspondance entre représentations irréductibles admissibles de GL(n,F) et représentations de degré n, à image semi-simple, de W_F pour $F = \mathbb{R}$ ou $\mathbb{C}$ (Langlands [6]; B. Speh [9]).

La suite exacte reliant les groupes de Weil :

$$1 \longrightarrow W_{\mathbb{C}} \longrightarrow W_{\mathbb{R}} \xrightarrow[\varepsilon]{} \{\pm 1\} \longrightarrow 1$$

permet d'obtenir par composition certaines représentations $r_{\mathbb{C}}$ de $W_{\mathbb{C}}$ à partir de représentations $r_{\mathbb{R}}$ de $W_{\mathbb{R}}$. Dans ce cas, on dit que $r_{\mathbb{C}}$ relève $r_{\mathbb{R}}$.

Comme l'action de $\{\pm 1\}$ sur $W_{\mathbb{C}} = \mathbb{C}^{\times}$ est celle du groupe de Galois, on en déduit qu'une représentation de $W_{\mathbb{C}}$ relève une représentation de $W_{\mathbb{R}}$ si et seulement si elle est isomorphe à sa composée par $(z \longmapsto \bar{z})$.

Considérons en particulier le cas de la série principale. Une représentation irréductible de la série principale de $GL(n,\mathbb{C})$ est définie par n caractères (non nécessairement unitaires) $\mu_1,\dots,\mu_n$ de $\mathbb{C}^{\times}$. Deux représentations ainsi définies sont équivalentes si et seulement si les deux n-uples sont identiques modulo permutation. La représentation de $W_{\mathbb{C}} = \mathbb{C}^{\times}$ associée est la représentation de degré n, $\mu_1 \oplus \dots \oplus \mu_n$.

Soit $\pi_{\mathbb{C}}(\mu_1,\dots,\mu_n)$ cette représentation de $GL(n,\mathbb{C})$, $r_{\mathbb{C}}(\mu_1,\dots,\mu_n)$ celle de $W_{\mathbb{C}}$. Notons encore σ l'automorphisme $z \longmapsto \bar{z}$ de $W_{\mathbb{C}} = \mathbb{C}^{\times}$. On voit que $r_{\mathbb{C}}$ relève une représentation de $W_{\mathbb{R}}$, i.e. $r_{\mathbb{C}}$ est stable par σ, si et seulement si la condition suivante est vérifiée :

(*) Il existe $\tau \in \mathfrak{S}_n$ tel que

$$\mu_i \circ \sigma = \mu_{\tau(i)} \quad i = 1,\dots n.$$

Si $\mu = (\mu_1,\dots,\mu_n)$ est régulier $(\mu_i \neq \mu_j)$, ceci implique que τ est une involution puisque $\sigma^2 = 1$. On peut montrer en fait que pour tout μ, la condition (*) implique :

(**) Il existe une involution $\tau \in \mathfrak{S}_n$, $\tau^2 = 1$ telle que

$$\mu_i \circ \sigma = \mu_{\tau(i)} \quad i = 1,\dots n.$$

On ne s'intéresse ici qu'au cas où ceci est vrai pour $\tau = 1$; ainsi

$$\mu_i = \mu_i \circ \sigma \quad \mu_i(z) = \mu_i(\bar{z}) \qquad i = 1,\dots n.$$

Ceci revient donc à supposer les μ_i <u>non ramifiés</u> :
$\mu_i(z) = \nu_i(z\bar{z})$, ν_i un caractère de $R^{+\times}$.

Considérons d'autre part la suite exacte :

$$1 \longrightarrow \mathbb{U} \longrightarrow W_{\mathbb{R}} \xrightarrow[\varepsilon.\det]{} \mathbb{R}^{\times} \longrightarrow 1 \qquad \mathbb{U} = \{|z| = 1\}$$

où det est défini par la réalisation matricielle de $W_{\mathbb{R}}$, et ε (déjà utilisé p.1) est donné par

$$\begin{pmatrix} a & \\ & \bar{a} \end{pmatrix} \longmapsto 1 \qquad \begin{pmatrix} & b \\ -\bar{b} & \end{pmatrix} \longmapsto -1.$$

A toute famille $(\nu_1 \ldots \nu_n)$ de caractères de $\mathbb{R}^{\times}$, on associe donc par composition une représentation $\nu_1 \oplus \ldots \oplus \nu_n$ de $W_{\mathbb{R}}$, notée $r_{\mathbb{R}}(\nu_1 \ldots \nu_n)$.

On vérifie alors que si $\mu_i = \nu_i \circ N_{\mathbb{C}/\mathbb{R}}$ (i.e. $\mu_i(z) = \nu_i(z\bar{z})$) alors $r_{\mathbb{C}}(\mu_1, \ldots \mu_n)$ relève $r_{\mathbb{R}}(\nu_1, \ldots \nu_n)$.

Soit alors $\Pi_{\mathbb{R}}(\nu_1 \ldots \nu_n)$ la classe d'isomorphisme de représentations associée par Langlands à $r_{\mathbb{R}}(\nu_1, \ldots \nu_n)$[1]. Pour des ν_i "génériques", c'est l'induite à $GL(n,\mathbb{R})$ du caractère du sous-groupe de Borel défini de façon évidente par $(\nu_1, \ldots \nu_n)$. On dira que $\Pi_{\mathbb{C}}(\mu_1, \ldots \mu_n)$ <u>relève</u> $\Pi_{\mathbb{R}}(\nu_1, \ldots \nu_n)$. On peut alors se demander comment traduire cette relation au niveau des représentations admissibles $\Pi_{\mathbb{C}}$ et $\Pi_{\mathbb{R}}$ des groupes $GL(n)$.

Pour $GL(2)$, la réponse a été donnée par Saito et Shintani ([7],[8]). Voici comment on peut la formuler dans le cas "générique", i.e. quand $\Pi_{\mathbb{C}}(\mu_1, \ldots \mu_n)$ et $\Pi_{\mathbb{R}}(\nu_1, \ldots \nu_n)$ sont des représentations de la série principale.

---

1. Il résulte des résultats obtenus par B. Speh dans sa thèse [9] que l'ensemble Π_r de classes de représentations L-indistingables de $GL(n,\mathbb{R})$ associée par Langlands à $r = r_{\mathbb{R}}(\nu_1, \ldots \nu_n)$ est bien réduit à un élément.

On peut définir une application norme N, analogue à $N_{\mathbb{C}/\mathbb{R}}$, de $GL(2,\mathbb{C})$ dans l'ensemble des classes de conjugaison de $GL(2,\mathbb{R})$. En particulier, si χ est une fonction invariante par conjugaison sur $G_{\mathbb{R}}$, $\chi \circ N$ est une fonction bien définie sur $G_{\mathbb{C}}$.

Remarquons maintenant que si $r_{\mathbb{C}}$ est stable par $(z \longmapsto \bar{z})$, la représentation $\pi_{\mathbb{C}}$ est stable par σ, i.e. $\pi_{\mathbb{C}}$ et $\pi_{\mathbb{C}} \circ \sigma$ sont isomorphes. Soit alors A un automorphisme de l'espace de la représentation tel que $\pi_{\mathbb{C}}(\sigma g) = A\pi_{\mathbb{C}}(g) A^{-1}$. Comme $\pi_{\mathbb{C}}$ est irréductible, le lemme de Schur montre qu'on peut choisir A tel que $A^2 = 1$. On définit, de façon analogue aux caractères des représentations irréductibles, le "caractère tordu"
$tr(\pi_{\mathbb{C}}(g)(A)$; c'est une distribution sur $G_{\mathbb{C}}$ et, au moins si $\pi_{\mathbb{C}}$ est associée à des caractères non ramifiés de $\mathbb{C}^{\times}$, une fonction localement sommable, analytique sur un certain ouvert $G''_{\mathbb{C}}$ (voir §.2).

Plus généralement, soit $\rho_{\mathbb{R}}(\nu_1, \nu_2)$ et $\rho_{\mathbb{C}}(\mu_1, \mu_2)$ les représentations des séries principales (irréductibles ou non) de $GL(2,\mathbb{R})$ et $GL(2,\mathbb{C})$ associées à deux caractères de $\mathbb{R}^{\times}$ ou $\mathbb{C}^{\times}$. $\pi_F(\mu_1, \mu_2)$ coïncide avec $\rho_F(\mu_1, \mu_2)$ quand celle-ci est irréductible. Si μ_1, μ_2 sont des caractères non ramifiés de $\mathbb{C}^{\times}$, on peut encore construire un opérateur involutif A, entrelaçant $\rho_{\mathbb{C}} = \rho_{\mathbb{C}}(\mu_1, \mu_2)$ et $\rho_{\mathbb{C}} \circ \sigma$ (voir §.3).
L'opérateur A étant choisi comme au §.3, on a le théorème suivant [1] :

Théorème 1 (Shintani [8])

$G_{\mathbb{C}} = GL(2,\mathbb{C}) \quad G_{\mathbb{R}} = GL(2,\mathbb{R})$

$\rho_{\mathbb{C}} = \rho_{\mathbb{C}}(\mu_1, \mu_2)$ $\rho_{\mathbb{R}} = \rho_{\mathbb{R}}(\nu_1, \nu_2)$ représentations de la série principale (irréductible ou non) ; μ_i non ramifiés.

Alors on a équivalence de

(i) Il existe $\varepsilon = \pm 1$ tel que $tr(\pi_{\mathbb{C}}(g)A) = \varepsilon \, tr\, \pi_{\mathbb{R}}(Ng)$ pour $g \in G''_{\mathbb{C}}$

(ii) $(\mu_1, \mu_2) = (\nu_1 \circ N_{\mathbb{C}/\mathbb{R}}, \nu_2 \circ N_{\mathbb{C}/\mathbb{R}})$ mod. permutation

1. Shintani décrit aussi le relèvement de la série discrète de $GL(2,\mathbb{R})$ ce qui correspond (p.4) à τ d'ordre 2.

On vérifie (cf. ci-dessous) qu'en termes de dualité de Langlands ceci implique :

Théorème 2

Soit $\Pi_{\mathbb{C}} = \Pi_{\mathbb{C}}(\mu_1, \mu_2)$ une représentation de la série principale irréductible de GL(2,$\mathbb{C}$). Alors $\Pi_{\mathbb{C}}$ relève une représentation admissible irréductible $\Pi_{\mathbb{R}}$ de GL(2,$\mathbb{R}$) si et seulement si il existe $\varepsilon = \pm 1$ tel que :

$$\mathrm{tr}(\Pi_{\mathbb{C}}(g)A) = \varepsilon \, \mathrm{tr}\, \Pi_{\mathbb{R}}(Ng) \qquad g \in G''_{\mathbb{C}}.$$

On va démontrer un théorème analogue au théorème 1, dans le cas de GL(n). Soit donc $G_{\mathbb{C}} = GL(n,\mathbb{C})$, $G_{\mathbb{R}} = GL(n,\mathbb{R})$. On peut définir une application norme de $G''_{\mathbb{C}}$ dans l'ensemble des classes de conjugaison de $G_{\mathbb{R}}$, $G''_{\mathbb{C}}$ étant un ouvert dense de $G_{\mathbb{C}}$.
Pour une représentation σ-stable $\rho_{\mathbb{C}} = \rho_{\mathbb{C}}(\mu_1, \dots \mu_n)$ de la série principale de GL(n,$\mathbb{C}$), avec les μ_i non ramifiés, on définit encore un opérateur d'entrelacement involutif A : $\rho_{\mathbb{C}} \sim \rho_{\mathbb{C}} \circ \sigma$; on peut ainsi définir le caractère tordu $\mathrm{tr}(\Pi_{\mathbb{C}}(g)A)$. C'est une fonction analytique sur $G''_{\mathbb{C}}$, localement sommable sur $G_{\mathbb{C}}$. Le théorème 1 est alors valable point par point pour GL(n)[1].

La démonstration est géométrique et utilise le théorème d'Atiyah-Bott. On sait qu'une représentation de la série principale est donnée par l'action de G sur les sections d'un fibré homogène sur G/B, B un sous-groupe de Borel. Le théorème d'Atiyah-Bott permet de calculer la valeur du caractère en un point régulier g de G, à partir de l'action de g sur le fibré au voisinage des points fixes de g sur G/B. On peut comparer ces actions, pour $G_{\mathbb{C}}/B_{\mathbb{C}}$ et $G_{\mathbb{R}}/B_{\mathbb{R}}$ respectivement, et en déduire l'identité entre caractère et caractère tordu.

Remarquons que cette méthode n'utilise pas de propriétés particulières de GL(n) et serait applicable à n'importe quelle forme réelle

---

1. Voir §.3 pour la formulation explicite du théorème.

déployée d'un groupe (semi-simple, réductif) complexe. On s'est contenté ici de donner la démonstration dans le cas de GL(n) car la définition de l'application norme serait plus problématique en général.

Comme pour GL(2), donnons les conséquences de ce théorème en termes de dualité de Langlands. On se limite ici au cas des paramètres réguliers. Vogan-Speh [10] donnent un critère de réductibilité qui, pour la série principale de $GL(n,\mathbb{R})$, se traduit ainsi (on note ν^+ la restriction à $\mathbb{R}^{+\times}$ d'un caractère ν de $\mathbb{R}^\times$).

> Soit $\rho_{\mathbb{R}}(\nu_1,\dots \nu_n)$, $\nu_i^+ \neq \nu_j^+$ pour $i \neq j$, une représentation de la série principale de $GL(n,\mathbb{R})$.
>
> $\rho_{\mathbb{R}}(\nu_1,\dots \nu_n)$ est <u>réductible</u> si et seulement si
>
> (*) il existe i,j tels que, pour un $n \in \mathbb{Z}^\times$
>
> $$\nu_i\nu_j^{-1}(x) = x^n \operatorname{sgn}(x),\ x \in \mathbb{R}^\times$$

D'autre part, on a des critères de réductibilité bien connus, dus à Želobenko et Wallach, pour les séries principales des groupes complexes (cf. Duflo [3]). Dans le cas d'une représentation non ramifiée de $GL(n,\mathbb{C})$:

> $\rho_{\mathbb{C}}(\mu_1,\dots \mu_n)$ (μ_i <u>non ramifiés</u>) est <u>réductible</u> ssi
>
> (**) il existe i,j, et $n \in \mathbb{Z}^\times$ tels que
>
> $$\mu_i\mu_j^{-1}(z) = |z|_{\mathbb{C}}^n \quad z \in \mathbb{C}^\times$$

Ici $|z|_{\mathbb{C}}$ est le module complexe : $|z|_{\mathbb{C}} = z\bar{z}$.

Soit alors $\pi_{\mathbb{C}} = \pi_{\mathbb{C}}(\mu_1,\dots \mu_n)$ une représentation de la série principale irréductible de $GL(n,\mathbb{C})$, non ramifiée (i.e. μ_i non ramifiés). On suppose de plus $\mu_i \neq \mu_j$ pour $i \neq j$. La représentation $\pi_{\mathbb{C}}$ est isomorphe à $\pi_{\mathbb{C}} \circ \sigma$; soit A l'un des deux opérateurs involutifs qui réalisent cet isomorphisme.

Par définition $\pi_{\mathbb{C}}$ relève une représentation (irréductible admissible) $\pi_{\mathbb{R}}$ de $G_{\mathbb{R}}$ si et seulement si

$$\pi_{\mathbb{R}} = \pi_{\mathbb{R}}(\nu_1 \dots \nu_n),\ \mu_i = \nu_i \circ N_{\mathbb{C}/\mathbb{R}} \text{ modulo } \mathfrak{S}_n.$$

$\pi_{\mathbb{C}}$ étant irréductible, on a d'après (**) :

$$\mu_i \mu_j^{-1}(z) \neq |z|_{\mathbb{C}}^n \quad \text{pour tout } n \in \mathbb{Z}^{\times}$$

ce qui équivaut à $\nu_i \nu_j^{-1}(x) \neq x^n \quad x \in \mathbb{R}^{+\times}$.

Comme μ est régulier, $\mu_i \neq \mu_j$ et donc $\nu_i^+ \neq \nu_j^+$ pour $i \neq j$; le paramètre ν est donc régulier et on vérifie que la condition de réductibilité (*) est contredite. Donc $\rho_{\mathbb{R}}(\nu_1 \ldots \nu_n)$ est irréductible, i.e. $\pi_{\mathbb{R}} = \rho_{\mathbb{R}}(\nu_1, \ldots \nu_n)$. Le théorème de relèvement étant prouvé pour les séries principales, on en déduit qu'il existe $\varepsilon = \pm 1$ tel que

$$t_r(\pi_{\mathbb{C}}(g)A) = \varepsilon \operatorname{tr} \pi_{\mathbb{R}}(Ng) \qquad g \in G''_{\mathbb{C}}.$$

Réciproquement, soit $\pi^{\circ}_{\mathbb{R}}$ une représentation irréductible admissible de $G_{\mathbb{R}}$ vérifiant (pour un $\varepsilon = \pm 1$)

$$\operatorname{tr}(\pi_{\mathbb{C}}(g)A) = \varepsilon \operatorname{tr} \overset{\circ}{\pi}_{\mathbb{R}}(Ng) \qquad g \in G''_{\mathbb{C}}.$$

Nous allons montrer que $\pi^{\circ}_{\mathbb{R}}$ est relevée par $\pi_{\mathbb{C}}$.

Soit $\pi_{\mathbb{R}} = \pi_{\mathbb{R}}(\nu_1, \ldots \nu_n)$ l'une des représentations de $G_{\mathbb{R}}$ qui relève $\pi_{\mathbb{C}}$. Soit χ, χ° les caractères-distributions respectifs (au sens de Harish-Chandra) de $\pi_{\mathbb{R}}$ et $\pi^{\circ}_{\mathbb{R}}$:

$$\chi(g) = \operatorname{tr} \pi_{\mathbb{R}}(g), \quad \chi^{\circ}(g) = \operatorname{tr} \pi^{\circ}_{\mathbb{R}}(g) \quad g \in G'_{\mathbb{R}}.$$

L'image de $G''_{\mathbb{C}}$ par l'application norme est ouverte dans $G_{\mathbb{R}}$. D'après les identités vérifiées respectivement par χ et χ°, on voit que pour un $\varepsilon = \pm 1$:

$$\chi = \varepsilon \chi^{\circ} \text{ sur un ouvert de } G_{\mathbb{R}}.$$

Soit $U(\mathfrak{g}_{\mathbb{C}})$ l'algèbre enveloppante de l'algèbre de Lie complexifiée $\mathfrak{g}_{\mathbb{C}}$ de $G_{\mathbb{R}}$, $\mathfrak{z}$ son centre. On sait que χ et χ° sont des distributions propres pour $\mathfrak{z}$:

$$x.\chi = \lambda(x)\chi \qquad x.\chi^{\circ} = \lambda^{\circ}(x)\chi^{\circ} \quad x \in \mathfrak{z}$$

λ et λ° étant des caractères de $\mathfrak{z}$, les caractères infinitésimaux des représentations $\pi_{\mathbb{R}}$ et $\pi^{\circ}_{\mathbb{R}}$. Comme χ et χ° coïncident (à un scalaire près) sur un ouvert, on en déduit que λ et λ° coïncident.

D'autre part, d'après le théorème du sous-quotient d'Harish-Chandra, $\pi^{\circ}_{\mathbb{R}}$ se plonge dans une représentation $\rho^{\circ}_{\mathbb{R}}$ de la série principale. $\rho^{\circ}_{\mathbb{R}}$ doit avoir le même caractère infinitésimal que $\pi^{\circ}_{\mathbb{R}}$, et donc que $\pi_{\mathbb{R}}$. Il est classique que ceci implique

$$\rho^{\circ}_{\mathbb{R}} = \rho_{\mathbb{R}}(\nu^{\circ}_1, \dots \nu^{\circ}_n), \text{ avec } (\nu^{\circ}_i)^{+} = \nu^{+}_i \mod \mathfrak{S}_n .$$

L'hypothèse de régularité sur μ montre alors que $\rho^{\circ}_{\mathbb{R}}$ est irréductible et donc égale à $\pi^{\circ}_{\mathbb{R}}$. Donc $\pi^{\circ}_{\mathbb{R}}$ est de la forme $\pi_{\mathbb{R}}(\nu_1, \dots \nu_n)$ avec

$$\mu_i = \nu_i \circ N_{\mathbb{C}/\mathbb{R}} \text{ modulo } \mathfrak{S}_n \text{ : ainsi } \pi_{\mathbb{C}} \text{ relève } \pi^{\circ}_{\mathbb{R}}.$$

En définitive, on a démontré [1] :

<u>Théorème</u> 3

Soit $\pi_{\mathbb{C}} = \pi_{\mathbb{C}}(\mu_1, \dots \mu_n)$ une représentation de la série principale irréductible de $GL(n,\mathbb{C})$. On suppose de plus le paramètre <u>régulier</u> : $\mu_i \neq \mu_j$ pour $i \neq j$.

Alors $\pi_{\mathbb{C}}$ relève une représentation irréductible admissible $\pi_{\mathbb{R}}$ de $GL(n,\mathbb{R})$ si et seulement si il existe $\varepsilon = \pm 1$ tel que :

$$t_r(\pi_{\mathbb{C}}(g)A) = \varepsilon \operatorname{tr} \pi_{\mathbb{R}}(Ng) \qquad g \in G''_{\mathbb{C}}.$$

---

1. Dans le cas de GL(2), les résultats plus complets connus sur la réductibilité (cf. Godement [4]) permettent de se passer de l'hypothèse de régularité.

2 - L'application norme

On démontre ici, de façon élémentaire, les résultats dont nous aurons besoin concernant l'application norme de $G_{\mathbb{C}}$ dans $G_{\mathbb{R}}$. Les démonstrations s'inspirent de Saito [7, §.3].

Soit $G_{\mathbb{C}} = GL(n,\mathbb{C})$, $G_{\mathbb{R}} = GL(n,\mathbb{R})$.

On désigne par σ ou par $(g \longmapsto \bar{g})$ l'involution de $G_{\mathbb{C}}$ par rapport à $G_{\mathbb{R}}$.

Définition

Soit $g,h \in G_{\mathbb{C}}$. On dit que g et h sont <u>σ-conjugués</u> si $g = x^{\sigma}hx^{-1}$ pour un $x \in G_{\mathbb{C}}$.

Dans $G_{\mathbb{C}}$ et $G_{\mathbb{R}}$, soit $G'_{\mathbb{C}}$ et $G'_{\mathbb{R}}$ les ouverts formés des éléments <u>réguliers</u>, c'est-à-dire des éléments (nécessairement semi-simples) dont les valeurs propres sont distinctes. Il est clair que $G'_{\mathbb{C}} \cap G_{\mathbb{R}} = G'_{\mathbb{R}}$.

Définition

On dit que $g \in G_{\mathbb{C}}$ est σ-régulier si $g\bar{g}$ est un élément régulier de $G_{\mathbb{C}}$.

Soit $G''_{\mathbb{C}}$ l'ensemble des éléments <u>σ-réguliers</u> de $G_{\mathbb{C}}$: c'est un ouvert dense, complémentaire d'un sous-ensemble analytique réel.

Soit $g \in G''_{\mathbb{C}}$. Alors $x = g\bar{g}$ est un élément régulier de $G_{\mathbb{C}}$. De plus x est conjugué à $\bar{x}$ dans $G_{\mathbb{C}}$: $\bar{g}g = g(\bar{g}g)g^{-1}$, donc si $P(x,T)$ est le polynôme caractéristique de x, $P(x,T) = P(\bar{x},T) = \bar{P}(x,T)$, i.e. $P(x,T)$ est <u>réel</u>, et il a ses racines distinctes puisque x est régulier. On en déduit que x est conjugué dans $G_{\mathbb{C}}$ à une matrice réelle y. Celle-ci est dans $G'_{\mathbb{C}} \cap G_{\mathbb{R}} = G'_{\mathbb{R}}$. De plus, deux matrices réelles, conjuguées dans $G_{\mathbb{C}}$, le sont dans $G_{\mathbb{R}}$, et on en déduit que y est défini à conjugaison près dans $G_{\mathbb{R}}$.

A un élément g de $G''_{\mathbb{C}}$, on a donc associé une classe de conjugaison régulière de $G_{\mathbb{R}}$. On la note Ng. Il est clair que si g,h sont σ-conjugués, Ng = Nh. On a donc défini une application

$$N : G''_{\mathbb{C}}/\sigma\text{-conjugaison} \longrightarrow G'_{\mathbb{R}}/\text{conjugaison}$$

Cherchons à caractériser l'image de N. Tout d'abord, $\det(g\bar{g}) = \det g(\overline{\det g}) > 0$. L'image de N est donc formée de classes de $GL^+(n,\mathbb{R})$. En fait :

<u>L'image de N est formée des classes de conjugaison (pour $G_{\mathbb{R}}$) de $G'_{\mathbb{R}} \cap (G_{\mathbb{R}})^2$.</u>

Il est clair que si $x \in G'_{\mathbb{R}}$ est un carré, la classe de x est dans l'image de N. Réciproquement, montrons que si la classe de $x \in G'_{\mathbb{R}}$ est une norme, x est un carré.

Soit $g = a + ib \in G''_{\mathbb{C}}$, a,b matrices réelles

$$\bar{g} = a - ib$$

$$g\bar{g} = a^2 + i(ba - ab) + b^2.$$

On a supposé $x = g\bar{g} \in G_{\mathbb{R}}$, donc :

$$g\bar{g} = a^2 + b^2 \text{ avec } ab - ba = [a,b] = 0 .$$

a,b réelles, $g\bar{g}$ semi-simple régulier.

Comme $g\bar{g}$ est semi-simple régulier dans $G_{\mathbb{R}}$, on peut le mettre (cf. fin du §) sous la forme

$$x = g\bar{g} = \begin{pmatrix} \lambda_1 & & & & & & & \\ & \ddots & & & & & & \\ & & \lambda_r & & & & & \\ & & & \mu_1 & & & & \\ & & & & \bar{\mu}_1 & & & \\ & & & & & \ddots & & \\ & & & & & & \mu_s & \\ & & & & & & & \bar{\mu}_s \end{pmatrix} = a^2 + b^2$$

dans une base de $\mathbb{C}^n$ de la forme $(f_1, \dots f_r, \ell_1, \bar{\ell}_1, \dots \ell_s, \bar{\ell}_s)$, f_i réels, $\lambda_i \in \mathbb{R}^\times$, $\mu_i \in \mathbb{C}^\times$.

Soit $\mathfrak{g}^x_{\mathbb{C}}$, $\mathfrak{g}^x_{\mathbb{R}}$ les commutants de x dans $\mathfrak{g}_{\mathbb{C}}$, $\mathfrak{g}_{\mathbb{R}}$, les algèbres de Lie de $G_{\mathbb{C}}$ et $G_{\mathbb{R}}$. Comme les valeurs propres $(\lambda_1, \dots, \bar{\mu}_s)$ sont distinctes, $\mathfrak{g}^x_{\mathbb{C}}$ s'identifie dans cette base à l'ensemble des matrices diagonales

$$y = \begin{pmatrix} y_1 & & 0 \\ & \ddots & \\ 0 & & y_{r+2s} \end{pmatrix} \qquad y_i \in \mathbb{C}$$

La conjugaison complexe opère dans $\mathcal{Y}_{\mathbb{C}}^{x}$ par

$$\sigma : y_i \longmapsto \bar{y}_i \quad i = 1,\dots r$$

$$(y_j, y_{j+1}) \longmapsto (\bar{y}_{j+1}, \bar{y}_j) \quad j = r+1,\ r+3,\dots$$

On en déduit que $\mathcal{Y}_{\mathbb{R}}^{x}$ est l'ensemble des matrices de la forme

$$(*) \qquad \begin{pmatrix} y'_1 & & & & & & & \\ & \ddots & & & & & & \\ & & y'_r & & & 0 & & \\ & & & y_{r+1} & & & & \\ & & & & \bar{y}_{r+1} & & & \\ & 0 & & & & y_{r+3} & & \\ & & & & & & \bar{y}_{r+3} & \\ & & & & & & & \ddots \end{pmatrix} \qquad \begin{matrix} y'_i \in \mathbb{R} \\ y_j \in \mathbb{C} \end{matrix}$$

Comme $x = a^2 + b^2$ et que a,b commutent, on en déduit que $a,b \in \mathcal{Y}_{\mathbb{R}}^{x}$. Soit $a = (a'_i, a_j)$, $b = (b'_i, b_j)$ la décomposition de a et b suivant (*). On a donc $\lambda_i = (a'_i)^2 + (b'_i)^2 > 0$, ce qui suffit à impliquer que x est un carré dans $GL(n,\mathbb{R})$.

□

On va maintenant prouver que N est une <u>injection</u> de l'ensemble des classes de σ-conjugaison de $G''_{\mathbb{C}}$ dans l'ensemble des classes de conjugaison de $G'_{\mathbb{R}}$. On doit donc montrer que si $Ng = Ng'$, g et g' sont σ-conjugués. Ceci se ramène à la propriété suivante :

> Soit $g_1, g_2 \in G''_{\mathbb{C}}$, tels que $g_1\bar{g}_1 = g_2\bar{g}_2$.
>
> Alors g_1 et g_2 sont σ-conjugués.

Soit $g_o = g_1\bar{g}_1 = g_2\bar{g}_2$. En transformant g_1 et g_2 par σ-conjugaison, on peut se ramener à $g_o \in G'_{\mathbb{R}}$. Soit Z_{g_o} le sous-groupe de Cartan, commutant de g_o dans $G_{\mathbb{C}}$. On a $g_i \in Z_{g_o}$ $(i=1,2)$. En effet $g_o = g_i\bar{g}_i = \bar{g}_i g_i$ puisque

$g_o = \bar{g}_o$ donc $g_i g_o = g_i \bar{g}_i g_i = g_o g_i$.

Z_{g_o} est <u>commutatif</u> puisque g_o est régulier, et stable par σ. On va démontrer le "théorème 90 de Hilbert" pour Z_{g_o} :

<u>Proposition</u> 1

Soit $g \in Z_{g_o}$ tel que $Ng = g\bar{g} = 1$.
Alors il existe $h \in Z_{g_o}$ tel que $g = h\bar{h}^{-1}$.

Montrons que ceci implique l'injectivité de N. Puisque $g_i \in Z_{g_o}$, les g_i commutent, et

$$1 = g_o^{-1} g_o = \bar{g}_1^{-1} g_1^{-1} g_2 \bar{g}_2 = (g_1^{-1} g_2)(\overline{g_1^{-1} g_2})$$

donc $N(g_1^{-1} g_2) = 1$, d'où $g_1^{-1} g_2 = h\bar{h}^{-1}$, $h \in Z_{g_o}$.

Donc $g_1 = hg_2\bar{h}^{-1}$ puisque tout commute, donc g_1 et g_2 sont σ-conjugués. On a montré :

<u>Proposition</u> 2

L'application N est une bijection des classes de $G''_{\mathbb{C}}$ pour la σ-conjugaison, sur les classes de carrés de $G'_{\mathbb{R}}$ pour la conjugaison

<u>Corollaire</u>

Tout élément de $G''_{\mathbb{C}}$ est σ-conjugué à un élément de $G_{\mathbb{R}}$.

On voit donc que $G''_{\mathbb{C}}$ n'est autre que l'ensemble des matrices σ-conjuguées à des matrices réelles de carré régulier.

Il reste à prouver la <u>Prop. 1</u>. Il s'agit de vérifier la nullité de la cohomologie en degré 1 du groupe de Galois $\mathfrak{G} = \{1,\sigma\}$ dans Z_{g_o}.

Comme g_o est semi-simple régulier, il admet une diagonalisation de la forme

$$\begin{pmatrix} \lambda_1 & & & & & & & \\ & \ddots & & & & & 0 & \\ & & \lambda_r & & & & & \\ & & & \mu_1 & & & & \\ & & & & \bar{\mu}_1 & & & \\ & & & & & \ddots & & \\ & 0 & & & & & \mu_s & \\ & & & & & & & \bar{\mu}_s \end{pmatrix}$$

avec les λ_i, μ_j, $\bar{\mu}_j$ tous distincts, dans une base de $\mathbb{C}^n$ de la forme $(\ell_1,\ldots \ell_r, f_1, \bar{f}_1,\ldots f_s, \bar{f}_s)$; $e_i \in \mathbb{R}^n$, $\bar{f}_j$ conjugué de f_j par rapport à $\mathbb{R}^n$. On en déduit que Z_{g_o} est isomorphe comme σ-module à $(V_1)^r \oplus (V_2)^s$, où

$V_1 \cong \mathbb{C}^\times$ avec l'action habituelle de σ :

$$\sigma z = \bar{z}$$

$V_2 \cong \mathbb{C}^\times \oplus \mathbb{C}^\times$ avec $\sigma(x,y) = (\bar{y},\bar{x})$.

Il suffit de vérifier la nullité de $H^1(\sigma, V_i)$.

Pour V_1 : c'est le théorème de Hilbert pour $\mathbb{C}^\times$.

Pour V_2 : soit $z = (x,y)$ tel que

$$zz^\sigma = (x,y)\ (\bar{y},\bar{x}) = (1,1).$$

Alors $x\bar{y} = 1$ et $z = (x,\bar{x}^{-1}) = (x,1)(1,\bar{x}^{-1}) = \zeta.(\zeta^\sigma)^{-1}$ avec $\zeta = (x,1)$.

3 - Le relèvement pour GL(n)

On va démontrer un théorème de relèvement pour toute la série principale, irréductible ou non.

Dans $GL(n,F) = G_F$ ($F = \mathbb{R}$ ou $\mathbb{C}$), on considère le sous-groupe B_F formé des matrices triangulaires supérieures. Si $\mu_1, \dots \mu_n$ sont n caractères (non nécessairement unitaires) de $F^\times$,

$$b \longmapsto \mu_1(x_1) \dots \mu_n(x_n) = \mu'(b), \quad b = \begin{pmatrix} x_1 & & * \\ & \ddots & \\ 0 & & x_n \end{pmatrix}$$

est un caractère de B_F. A un tel caractère est associé un G_F-fibré homogène complexe en droites, $E'_{\mu'}$, sur la variété compacte $X_F = G_F/B_F$. X_F étant compacte, on peut définir sans ambiguïté l'espace des sections L^2 de ce fibré. G_F opère sur l'espace de Hilbert des sections L^2, définissant une représentation de la série principale ; la représentation associée sur l'espace des vecteurs C^∞ est donnée par l'action de G_F sur les sections C^∞ du fibré.

Si $\mu'(b) = \mu_1(x_1) \dots \mu_n(x_n)\, \delta_F(b)^{1/2}$, où

$\delta_F(b) = \prod_{i=1}^{n} |x_i|_F^{n+1-2i}$ est le module de B_F, la représentation ainsi associée à $\mu = (\mu_1, \dots \mu_n)$ sera notée $\rho_F(\mu_1, \dots \mu_n)$ ou $\rho_F(\mu)$. (Ici $|x|_{\mathbb{R}} = |x|$ pour $x \in \mathbb{R}$, $|x|_{\mathbb{C}} = x\bar{x}$ pour $x \in \mathbb{C}$). Le fibré $E'_{\mu'}$ associé sera noté E_μ. On obtient ainsi la paramétrisation habituelle de la série principale ; en particulier, $\rho_F(\mu)$ est unitaire quand les μ_i le sont.

Soit $\rho_{\mathbb{C}}(\mu_1, \dots \mu_n)$ une représentation de $G_{\mathbb{C}}$, et supposons les μ_i non ramifiés : $\mu_i(z) = \nu_i(z\bar{z})$, ν_i étant un caractère de $\mathbb{R}^\times$ défini à torsion près par Sgn. Si $\rho_{\mathbb{C}}$ est irréductible, la considération de son caractère montre alors qu'elle est infinitésimalement équivalente à sa composée avec σ. On va réaliser géométriquement un opérateur d'entrelacement entre $\rho_{\mathbb{C}}(\mu)$ et $\rho_{\mathbb{C}}(\mu) \circ \sigma$, pour tout μ non ramifié.

L'espace $\Gamma^\infty E_\mu$ des sections de E_μ s'identifie à l'espace des fonctions C^∞ sur $G_{\mathbb{C}}$ vérifiant

$$f\left[g\begin{pmatrix} x_1 & & * \\ & \ddots & \\ 0 & & x_n \end{pmatrix}\right] = \mu_1(x_1)\ldots\ \mu_n(x_n)\,\delta_{\mathbb{C}}^{1/2}\, f[g].$$

Soit $G_{\mathbb{C}} = K_{\mathbb{C}}B_{\mathbb{C}}$, $K_{\mathbb{C}} = U(n)$, la décomposition d'Iwasawa de $G_{\mathbb{C}}$. Soit $M_{\mathbb{C}} = K_{\mathbb{C}} \cap B_{\mathbb{C}}$: $M_{\mathbb{C}}$ est l'ensemble des matrices $\begin{pmatrix} x_1 & & 0 \\ & \ddots & \\ 0 & & x_n \end{pmatrix}$, $|x_i| = 1$.

Comme μ, $\delta_{\mathbb{C}}$ sont non ramifiés, $\mu\cdot\delta_{\mathbb{C}}\big|_{M_{\mathbb{C}}}$ est trivial et $\Gamma^\infty E_\mu$ s'identifie donc à l'espace des fonctions sur $K_{\mathbb{C}}$ vérifiant

$$\varphi\left[k\begin{pmatrix} x_1 & & 0 \\ & \ddots & \\ 0 & & x_n \end{pmatrix}\right] = \varphi(k),$$ i.e. à l'espace $C^\infty(K_{\mathbb{C}})$ des fonctions sur $K_{\mathbb{C}}/M_{\mathbb{C}}$ $\cong G_{\mathbb{C}}/B_{\mathbb{C}} = X_{\mathbb{C}}$. Autrement dit, comme $K_{\mathbb{C}}$-fibré homogène, E_μ est <u>trivial</u> et on vient d'en décrire une trivialisation. Le groupe $G_{\mathbb{C}}$ opère sur $C^\infty(X_{\mathbb{C}})$ par

$$\rho_{\mathbb{C}}(g)f(k\,M_{\mathbb{C}}) = f(\underline{k}(g^{-1}k).M_{\mathbb{C}})\ \mu'(\underline{b}(g^{-1}k))^{-1}$$

où on a noté $g = \underline{k}(g)\,\underline{b}(g)$ une décomposition arbitraire de $g \in G_{\mathbb{C}}$ suivant $G_{\mathbb{C}} = K_{\mathbb{C}}\,B_{\mathbb{C}}$.

On note σ ou $(k \longmapsto \bar{k})$ la conjugaison complexe dans $K_{\mathbb{C}}$, induite de celle de $G_{\mathbb{C}}$. L'automorphisme σ laisse stable $M_{\mathbb{C}}$ et induit donc une involution, aussi notée σ, de $X_{\mathbb{C}}$. L'ensemble des points fixes de σ dans $K_{\mathbb{C}}$ est $K_{\mathbb{R}} = O(n)$. On voit facilement que l'ensemble des points fixes de σ dans $X_{\mathbb{C}}$ est $X_{\mathbb{R}} = K_{\mathbb{R}}/M_{\mathbb{R}} = G_{\mathbb{R}}/B_{\mathbb{R}}$, où $M_{\mathbb{R}} = K_{\mathbb{R}} \cap B_{\mathbb{R}}$ est formé des matrices diagonales à éléments (± 1).

<u>Lemme</u>

l'opérateur A: $L^2(X_{\mathbb{C}}) \longrightarrow L^2(X_{\mathbb{C}})$

$$f(x) \longmapsto f(\sigma x)$$

est un opérateur d'entrelacement involutif de $\rho_{\mathbb{C}}$ et $\rho_{\mathbb{C}} \circ \sigma$.

En effet :

$$[\varphi_{\mathbb{C}}(g)A]f(k\,M_{\mathbb{C}}) = Af(\underline{k}(g^{-1}k)M_{\mathbb{C}})\,\mu'(\underline{b}(g^{-1}k))^{-1}$$

$$= f(\sigma k(g^{-1}k)M_{\mathbb{C}})\,\mu'(\underline{b}(g^{-1}k))^{-1} \qquad (*)$$

$$[A\rho_{\mathbb{C}}\cdot\sigma(g)]f(k\,M_{\mathbb{C}}) = (\pi_{\mathbb{C}}(\sigma g)f)(\sigma k\,M_{\mathbb{C}})$$

$$= f(\underline{k}(\sigma g^{-1}.\sigma k)M_{\mathbb{C}})\,\mu'(\underline{b}(\sigma g^{-1}.\sigma k))^{-1} \qquad (**)$$

Or, $K_{\mathbb{C}}$ et $B_{\mathbb{C}}$ étant stables par σ,

$\underline{k}(\sigma g^{-1}\sigma k) = \sigma\underline{k}(g^{-1}k)$ et $\underline{b}(\sigma g^{-1}\sigma k) = \sigma\underline{b}(g^{-1}k)$;

μ' étant non ramifié, $\mu'(\sigma b) = \mu'(b)$ et on vérifie alors que les expressions (*) et (**) coïncident.

□

Lorsque $\rho_{\mathbb{C}}$ est irréductible, on sait qu'elle est topologiquement complétement irréductible ; d'après le lemme de Schur A est donc, au signe près, l'unique opérateur d'entrelacement involutif (continu $L^2(X_{\mathbb{C}}) \longrightarrow L^2(X_{\mathbb{C}})$) de $\rho_{\mathbb{C}}$ et $\rho_{\mathbb{C}}\cdot\sigma$.

Nous voulons maintenant définir "$\mathrm{tr}(\rho_{\mathbb{C}}(g)A)$". Nous utilisons les méthodes de Guillemin-Sternberg [5, ch.6].

Soit $E \longrightarrow X$ un fibré vectoriel sur une variété compacte X. Un endomorphisme (f,r) du fibré E est la donnée d'une application différentiable $f : X \longrightarrow X$ et d'une section différentiable r de $\underline{\mathrm{Hom}}(f^*E,E)$, i.e. d'une famille différentiable d'applications $r(x) : E_{f(x)} \longrightarrow E_x$. Un morphisme définit une application $\Gamma^\infty(X,E) \longrightarrow \Gamma^\infty(X,E)$ par

$$s \longmapsto t : t(x) = r(x).s(f(x)).$$

Par exemple, si $E \longrightarrow X$ est un G-fibré homogène, on a pour tout $g \in G$ une application $r_g : E \longrightarrow E$ telle que $r_g.E_x = E_{g.x}$, donc $r_g \in \mathrm{Hom}(t_g^*E,E)$ où $t_g x = g^{-1}.x$. Donc l'action de G définit bien un morphisme au sens ci-dessus.

Plus généralement, si Y est une variété (pas nécessairement un groupe), on dit que Y <u>agit transitivement sur</u> E si

1. On a une famille (f_y,r_y) de morphismes de E paramétrée par Y, et C^∞ en y dans un sens évident.

2. L'action de Y sur la base est localement transitive : posons $f(y,x) = f_y(x)$ $y \in Y$, $x \in X$. On suppose alors que l'application

$$T_yY \longrightarrow T_{f(y,x)}X$$
$$\eta \longrightarrow df_{(y,x)}(\eta,0)$$

est surjective pour tout (y,x).

Ces conditions sont évidemment satisfaites dans le cas d'un G-fibré homogène. Si $Y = G_{\mathbb{C}}$, et E_μ est le fibré défini ci-dessus pour $\mu = (\mu_1,\dots \mu_n)$ non ramifié, on obtient une action de $G_{\mathbb{C}}$ sur E_μ de la façon suivante :

– $f_g : X_{\mathbb{C}} \longrightarrow X_{\mathbb{C}}$ est le composé $t_g \circ \sigma$, où t_g est l'action de $g \in G_{\mathbb{C}}$ sur la base : $t_g.x = g^{-1}x$.

– On a fixé plus haut une trivialisation de E_μ ; r_g est alors donné par

$$E_{g^{-1}.\sigma x} \xrightarrow[\text{sur le fibré homogène}]{\text{action de } g} E_{\sigma x} \xrightarrow[\text{isomorphisme donné par la trivialisation}]{\approx} E_x .$$

(Ceci n'est pas une action de <u>groupe</u> de $G_{\mathbb{C}}$)[1].

Il est clair que l'action sur la base est localement transitive – et transitive ; entre les sections, (f_g, r_g) induit l'application $\rho_{\mathbb{C}}(g)A : \Gamma^\infty E_\mu \longrightarrow \Gamma^\infty E_\mu$.

On peut maintenant expliciter le théorème d'Atiyah-Bott (cf. Guillemin-Sternberg [5, p. 315]). Disons que $y \in Y$ opère régulièrement sur X si l'application f_y est transverse à l'identité, c'est-à-dire si son graphe $\Gamma \subset X \times X$ est transverse à la diagonale. En particulier, f_y n'a alors qu'un nombre fini de points fixes puisque X est compacte.

---

1. Mais on pourrait se ramener à une action de groupe de $G_{\mathbb{C}} \ltimes (1,\sigma)$.

Théorème 1 (Atiyah-Bott)

Soit (f,r) une action de Y, localement transitive, sur le fibré vectoriel à base compacte $E \longrightarrow X$.

1. Pour toute densité ρ, C^∞ à support compact sur Y, l'opérateur

$$\Pi_\rho = \int_Y \rho(y) f_y^* \quad : L^2E \longrightarrow L^2E$$

est compact, traçable. L'application $\rho \longrightarrow \operatorname{tr} \Pi_\rho$ définit une fonction généralisée sur Y, notée tr f^*.

2. Si $y \in Y$ opère régulièrement sur X, tr f^* est une fonction C^∞ au voisinage de y, et

$$(\operatorname{tr} f^*)(y) = \sum_{f_y(x)=x} \frac{\operatorname{tr} r_y(x)}{|\det(1-df_y(x))|}$$

(Ici L^2E est l'espace des sections L^2 du fibré, défini à l'aide d'une structure hermitienne sur E et d'une densité sur X, différentiables. L'espace L^2 ne dépend pas de ces choix parce que X est compacte).

Nous appliquons ceci aux deux situations suivantes :

- l'action tordue de $G_{\mathbb{C}}$ sur E_μ définie ci-dessus.

- l'action de $G_{\mathbb{R}}$ sur E_ν, où E_ν est le fibré homogène sur $X_{\mathbb{R}}$ défini par $\nu = (\nu_1, \ldots \nu_n)$.

La partie 2. du théorème 1 nous permet d'associer un sens aux expressions $\operatorname{tr}(\rho_{\mathbb{R}}(g))$, $\operatorname{tr}(\rho_{\mathbb{C}}(g)A)$ lorsque $g \in G_{\mathbb{R}}$ ou $G_{\mathbb{C}}$ opère régulièrement pour ces deux actions respectivement. On peut alors énoncer le théorème de relèvement :

Théorème 2

Soit $\mu = (\mu_1, \ldots \mu_n)$ des caractères non ramifiés de $\mathbb{C}^\times$. Soit $\rho_{\mathbb{C}}(\mu)$ la représentation de $G_{\mathbb{C}}$ associée. Soit A l'opérateur défini ci-dessus, qui entrelace $\rho_{\mathbb{C}}$ et $\rho_{\mathbb{C}} \circ \sigma$; on a $A^2 = 1$.

1. Pour tout $\varphi \in C_c^\infty(G_{\mathbb{C}})$, l'opérateur

$$\rho_{\mathbb{C}}(\varphi)A = \int_{G_{\mathbb{C}}} \rho_{\mathbb{C}}(g)A \ \varphi(g)dg$$

où dg est une mesure de Haar, est traçable. L'application $\varphi \longmapsto tr(\rho_{\mathbb{C}}(\varphi)A)$ définit une distribution sur $G_{\mathbb{C}}$, qui est une fonction C^{∞} au voisinage des éléments σ-réguliers. Soit $tr(\rho_{\mathbb{C}}(g)A)$ cette fonction C^{∞} sur $G''_{\mathbb{C}}$.

2. Pour $g \in G''_{\mathbb{C}}$,

$$(*) \qquad tr(\rho_{\mathbb{C}}(g)A) = tr\ \rho_{\mathbb{R}}(Ng)$$

pour toute représentation $\rho_{\mathbb{R}} = \rho_{\mathbb{R}}(\nu_1,\dots,\nu_n)$ de $G_{\mathbb{R}}$ telle que $\mu_i = \nu_i \circ N_{\mathbb{C}/\mathbb{R}}$ modulo permutation.

Dans l'égalité (*), $tr\ \rho_{\mathbb{R}}(Ng)$ désigne la valeur du caractère-distribution de $\rho_{\mathbb{R}}$ au point régulier Ng.

Démonstration

Il nous reste à montrer que pour g σ-régulier, l'action tordue de g sur $X_{\mathbb{C}}$ est régulière, et à établir l'égalité (*).

Considérons d'abord l'action naturelle de G_F sur $X_F = G_F/B_F$ ($F=\mathbb{R}$ ou $\mathbb{C}$). Il est clair que $g \in G_F$ a un point fixe dans X_F si et seulement si g est conjugué à un élément de B_F. Soit $g \in B_F$, g régulier. Modulo conjugaison, on peut supposer que $g \in H'_F = H_F \cap G'_F$,

$$H_F = \left\{ \begin{pmatrix} x_1 & & 0 \\ & \ddots & \\ 0 & & x_n \end{pmatrix} \right\}.$$

Lemme

h $\in H'_F$ a exactement $n! = |W|$ points fixes dans X_F ; ce sont les $w\,B_F$ pour $w \in W$.

W est le groupe de Weyl $N(H_F)/_{H_F} \cong \mathfrak{S}_n$, qui s'identifie à l'ensemble des matrices de permutation.

Démonstration

Il est clair que $w\,B_F$ est un point fixe :

$hw\,B_F = w.w^{-1}hw\,B_F = w.h^w\,B_F = w\,B_F$.

Réciproquement, soit $x = g\,B_F$ un point fixe :

$hg\,B_F = g\,B_F$ donc $g^{-1}hg \in B_F$. Mais alors $b = g^{-1}hg$ est un élément régulier de B_F, et il est donc conjugué dans B_F à un élément de H'_F : $b = yh'\,y^{-1}$, $h' \in H'_F$, $y \in B_F$ [1].

Si $z = gy$, on a alors $z^{-1}hz = h'$ d'où (h régulier) $z^{-1}\,H_F z = H_F$ donc $z \in N(H_F) = W\,H_F$, d'où $g \in W\,B_F$. $\square$

De plus, on vérifie facilement que si $g \in G'_F$ a des points fixes, l'action tangente aux points fixes est bien transverse à l'identité (voir ci-dessous).

Soit maintenant $g \in G''_{\mathbb{C}}$. On considère l'automorphisme $x \longmapsto g^{-1}.\sigma x$ de $X_{\mathbb{C}}$. Il est facile de voir que si g,h sont σ-conjugués, leurs actions sur $X_{\mathbb{C}}$ sont conjuguées par un difféomorphisme de $X_{\mathbb{C}}$; par conséquent, d'après les résultats du §.2, il suffit d'étudier la régularité de $g \in G''_{\mathbb{C}} \cap G_{\mathbb{R}}$, i.e. $g \in G_{\mathbb{R}}$, $g^2 \in G'_{\mathbb{R}}$. Il suffira pour établir (*) de calculer $tr(\rho_{\mathbb{C}}(g)A)$ pour g réel : on voit facilement que cette distribution est invariante par σ-conjugaison.

Soit donc $g \in G''_{\mathbb{C}} \cap G_{\mathbb{R}}$, et $x \in X_{\mathbb{C}}$ un point fixe de $g^{-1}\sigma$: $g^{-1}\sigma x = x$ d'où $g^{-1}\ \sigma g^{-1}\sigma x = x$.

Mais g étant réel, son action commute avec la conjugaison complexe, d'où $g^{-2}x = x$. Un point fixe de $g^{-1}\sigma$ dans $X_{\mathbb{C}}$ est donc point fixe de g^{-2}. Mais g^{-2} est régulier (dans $G_{\mathbb{R}}$ ou $G_{\mathbb{C}}$) donc il a n! points fixes dans $X_{\mathbb{R}}$, et de même dans $X_{\mathbb{C}}$: donc ses points fixes sont tous dans $X_{\mathbb{R}}$.

Donc tout point fixe de $g^{-1}\sigma$ est dans $X_{\mathbb{R}}$; comme $\sigma|_{X_{\mathbb{R}}} = id.$, on voit que les points fixes de $g^{-1}\sigma$ sont réels et sont ceux de g^{-1}.

---

1. Exercice : Si une matrice triangulaire a ses valeurs propres distinctes, on peut la diagonaliser à l'aide d'un changement de base triangulaire.

En particulier, $g^{-1}\sigma$ n'a de points fixes que pour g dans un conjugué de $B_{\mathbb{R}}$. On se ramène en utilisant la conjugaison (qui est la trace sur $G_{\mathbb{R}}$ de la σ-conjugaison) à considérer $g \in H_{\mathbb{R}} \cap G''_{\mathbb{C}}$:

$$g = \begin{pmatrix} x_1 & & 0 \\ & \ddots & \\ 0 & & x_n \end{pmatrix} \qquad x_i \in \mathbb{R}^{\times}, \quad |x_i| \neq |x_j|.$$

L'expression (2) du théorème d'Atiyah-Bott donne alors :

$$\operatorname{tr}(\rho_{\mathbb{C}}(g)A) = \sum_{w \in \mathfrak{S}_n} \frac{(\mu\delta_{\mathbb{C}}^{1/2})(x_{w.1},\ldots x_{w.n})}{|\det(1-d(g^{-1}\sigma))w\, B_{\mathbb{C}}|}$$

Soit $x = w\, B_{\mathbb{C}} \in X_{\mathbb{C}}$ un point fixe de $g^{-1}\sigma$. Soit $V_{\mathbb{R}}$ l'espace tangent en x à $X_{\mathbb{R}}$, $T_{\mathbb{R}} = d(g^{-1})_x$ l'endomorphisme tangent en x à l'action de g^{-1} sur $X_{\mathbb{R}}$. $V_{\mathbb{C}} = V_{\mathbb{R}} \otimes \mathbb{C}$ est l'espace tangent à $X_{\mathbb{C}}$, et la conjugaison complexe $J : V_{\mathbb{C}} \longrightarrow V_{\mathbb{C}}$ par rapport à $V_{\mathbb{R}}$ est l'endomorphisme tangent à σ. Le complexifié $T_{\mathbb{C}}$ de $T_{\mathbb{R}}$ est l'endomorphisme tangent à l'action de g^{-1} sur $X_{\mathbb{C}}$.

Selon la décomposition $V_{\mathbb{C}} = V_{\mathbb{R}} \oplus i\, V_{\mathbb{R}}$, on a donc

$$T_{\mathbb{C}} = \begin{pmatrix} T & \\ & T \end{pmatrix}, \quad T_{\mathbb{C}}J = \begin{pmatrix} T & \\ & -T \end{pmatrix}$$

et $1 - T_{\mathbb{C}}J = (1-d(g^{-1}\sigma))_{w\, B_{\mathbb{C}}} = \begin{pmatrix} 1-T & \\ & 1+T \end{pmatrix}$.

$$\det(1-T_{\mathbb{C}}J) = \det(1-T^2)$$

c'est-à-dire $|\det(1-d(g^{-1}\sigma))_{w\, B_{\mathbb{C}}}| = |\det(1-d(g^{-2}))_{w\, B_{\mathbb{R}}}|$.

La non-nullité du déterminant de gauche en tout point $w\, B_{\mathbb{C}}$ exprime que l'action de $g^{-1}\sigma$ est transverse à l'identité. La non-nullité du déterminant de droite exprime que l'action de g^{-2} sur $X_{\mathbb{R}}$ est transverse à l'identité. Un calcul simple montre que ce déterminant est égal à $\prod_{i<j}(1-x_i^2 x_j^{-2})$ (voir Guillemin-Sternberg [5, p. 318]). Il n'est pas nul puisque $|x_i| \neq |x_j|$ pour $i \neq j$, ce qui montre que l'action de g^{-2} et celle de $g^{-1}\sigma$ sont transverses à l'identité.

D'autre part on a donc :

$$\mathrm{tr}(\rho_{\mathbb{C}}(g)A) = \sum_{w \in \mathfrak{S}_n} \frac{(\mu\delta_{\mathbb{C}}^{1/2})(x_{w\cdot 1},\ldots x_{w\cdot n})}{|\det(1-d(g^{-2}))_{w\,B_{\mathbb{R}}}|}$$

$$g = \begin{pmatrix} x_1 & & 0 \\ & \ddots & \\ 0 & & x_n \end{pmatrix} \quad , \quad |x_i| \neq |x_j|.$$

En remarquant que $\delta_{\mathbb{C}}(x) = \delta_{\mathbb{R}}^2(x) = \delta_{\mathbb{R}}(x^2)$ pour $x \in B_{\mathbb{R}}$, le théorème d'Atiyah-Bott, appliqué cette fois à l'action naturelle de $G_{\mathbb{R}}$ sur E_{ν}, montre encore que ceci est égal à $\mathrm{tr}(\rho_{\mathbb{R}}(g^2))$, pour $\rho_{\mathbb{R}} = \rho_{\mathbb{R}}(\nu)$, si $\nu = (\nu_1,\ldots \nu_n)$ est choisi (modulo permutation) tel que

$$\nu_i(x^2) = \mu_i(x) \qquad x \in \mathbb{R}^{\times}$$

c'est-à-dire $\mu_i(z) = \mu_i(|z|) = \nu_i(|z|^2) = \nu_i(z\bar{z})$. Ceci définit ν_i, au signe près.

□

4 - Intégrabilité des caractères gauches

Bien que nous ne l'ayons pas utilisé dans ce qui précède, nous voulons indiquer ici que les caractères gauches (ou "tordus") de la série principale non ramifiée ont la propriété d'intégrabilité locale analogue à celle démontrée par Harish-Chandra pour tous les caractères irréductibles.

Proposition

> $\mu = (\mu_1, \dots \mu_n)$ caractères non ramifiés de $\mathbb{C}^\times$. $\rho_{\mathbb{C}} = \rho_{\mathbb{C}}(\mu)$ représentation de $G_{\mathbb{C}}$.
>
> A défini comme au §.3.
>
> Alors la distribution $\varphi \longmapsto tr(\rho_{\mathbb{C}}(\varphi)A)$ est donnée par une fonction localement intégrable sur $G_{\mathbb{C}}$, qui coïncide sur $G''_{\mathbb{C}}$ avec la fonction $tr(\rho_{\mathbb{C}}(g)A)$ définie au §.3 ; elle est donc C^∞ sur $G''_{\mathbb{C}}$.

On peut démontrer ceci de deux manières :

a. On peut exprimer $\rho_{\mathbb{C}}(\varphi)A$, $\varphi \in C_c^\infty(G_{\mathbb{C}})$, comme un opérateur intégral sur $L^2(K_{\mathbb{C}})$ Procédant comme dans le cas "non tordu" (voir e.g. Wallach [11 , p. 244-47]), on exprime alors sa trace, par intégration sur la diagonale, comme une intégrale de la forme

$$\int_{K_{\mathbb{C}}AN_{\mathbb{C}}} a^\lambda \varphi(k^\sigma an\, k^{-1})dk\, dn\, da$$

où $G_{\mathbb{C}} = K_{\mathbb{C}} AN_{\mathbb{C}}$, $A = \left\{ \begin{pmatrix} x_1 & & 0 \\ & \ddots & \\ 0 & & x_n \end{pmatrix} \right\}$ $x_i > 0$, $N_{\mathbb{C}} = \left\{ \begin{pmatrix} 1 & & \\ & \ddots & \\ 0 & & 1 \end{pmatrix} \right\}$, est la décomposition d'Iwasawa de $G_{\mathbb{C}}$, et $a \longmapsto a^\lambda$ est un certain caractère de A, déterminé par μ.

On transforme alors cette intégrale en une intégrale de la forme

$$\int_{G_{\mathbb{C}}} \varphi(g)\, \chi(g)dg$$

à l'aide de l'application $K_{\mathbb{C}}\, AN_{\mathbb{C}} \longrightarrow G_{\mathbb{C}}$

$$k,a,n \longmapsto k^\sigma ank^{-1}.$$

On vérifie que χ est une fonction localement sommable. (Pour ceci dans le cas GL(2) voir Clozel [2]).

b. On peut aussi utiliser un résultat général de Guillemin-Sternberg [5, p. 317]. Soit G un groupe de Lie, H un sous-groupe fermé tel que G/H soit compact, $E \longrightarrow X = G/H$ un G-fibré homogène. Soit ρ la représentation de G sur les sections L^2 de E. On définit (§.3) la distribution $\mathrm{tr}\,\rho$ sur G. Soit $\mathfrak{g}$, $\mathfrak{h}$ les algèbres de Lie de G; H.

Proposition (Guillemin-Sternberg)

S'il existe $\eta \in \mathfrak{h}$ tel que $\mathrm{ad}(\eta)$ induit un isomorphisme de $\mathfrak{g}/\mathfrak{h}$ sur lui-même, le caractère $\mathrm{tr}\,\rho$ est représenté par une fonction sommable sur G.

Il suffit d'appliquer cette proposition à la paire (G,H) où G est le produit semi-direct $G_{\mathbb{C}} \rtimes \{1,\sigma\}$ et $H = B_{\mathbb{C}} \rtimes \{1,\sigma\}$ - considérer la composante $G_{\mathbb{C}} \rtimes \sigma$ de G.

Références

1. M.F. Atiyah et R. Bott, "A Lefschetz fixed point formula for elliptic complexes I", Ann. of Math. (2) 86 (1967), 374-407.

2. L. Clozel, en préparation.

3. M. Duflo, "Représentations irréductibles des groupes semi-simples complexes", Lecture Notes in Math, n° 497, 1975, p. 26-88.

4. R. Godement, Notes on Jacquet-Langlands' theory, IAS 1970.

5. V. Guillemin et S. Sternberg, Geometric Asymptotics, Math. Surveys 14, AMS 1977.

6. R.P. Langlands, On the classification of irreducible representations of real algebraic groups, IAS 1973 (preprint).

7. H. Saito, Automorphic Forms and algebraic extensions of number fields, Lectures in Math. 8, Dept. of Math., Kyoto University, 1975.

8. T. Shintani, "On irreducible unitary characters of a certain group extension of GL(2,ℂ)", J. Math. Soc. Japan, 29, 1, 1977.

9. B. Speh, Some Results on principal series for GL(n,ℝ), thèse, MIT 1977.

10. D. Vogan, B. Speh, "Reducibility of generalized principal series representations", à paraître.

11. N. Wallach, Harmonic Analysis on homogeneous spaces, Marcel Dekker, New York 1973.

Sur la méthode des orbites

par J.Dixmier

Introduction.

Soient $\underline{g}$ une algèbre de Lie de dimension finie sur un corps algébriquement clos de caractéristique 0 , $U(\underline{g})$ son algèbre enveloppante, Prim $U(\underline{g})$ l'ensemble des idéaux primitifs de $U(\underline{g})$. Pour calculer Prim $U(\underline{g})$, on imite ce qu'a fait A.Kirillov pour les groupes de Lie : On tente d'associer à toute orbite ω du groupe adjoint algébrique dans le dual $\underline{g}^*$ de $\underline{g}$ un idéal primitif $I(\omega)$ de $U(\underline{g})$, et cela en utilisant les polarisations des éléments de ω . Cette méthode réussit pour $\underline{g}$ résoluble. Dans le cas général, on se heurte à plusieurs difficultés, entre autres les deux suivantes : 1) si $f \in \underline{g}^*$, il peut arriver que f soit non polarisable; 2) si f admet deux polarisations $\underline{p}_1, \underline{p}_2$, les idéaux construits à partir de $\underline{p}_1$ et $\underline{p}_2$ peuvent être distincts.

Etant donnée une orbite ω , il faudrait lui associer directement un élément de Prim $U(\underline{g})$, sans passer par les polarisations. Dans le cas des groupes de Lie nilpotents, le problème analogue a une solution, due à Kirillov : On considère la mesure de Kostant de ω , on prend sa transformée de Fourier dans $\underline{g}$, on envoie cette transformée dans le groupe par l'application exponentielle, et l'on obtient le caractère-distribution de la représentation associée à ω . Pour les algèbres enveloppantes et dans le cas non nilpotent, un analogue partiel a été obtenu par M.Duflo ([7],th.3). Ces méthodes sont non algébriques.

Dans ce mémoire, nous décrirons, pour les algèbres de Lie nilpotentes, une construction algébrique inspirée par la construction analytique de Kirillov-Duflo. Le résultat principal est le th.5.3. Au §6, nous verrons qu'il existe un vague espoir d'étendre cette construction au cas général; en particulier, nous décrirons un exemple où notre méthode fournit, à partir d'orbites <u>non polarisables</u>, des idéaux primitifs de $U(\underline{g})$.

Dans tout le mémoire, k désigne un corps commutatif de caractéristique 0 .

1. L'algèbre A(V) d'un espace vectoriel V .

1.1. Soit V un espace vectoriel de dimension finie sur k . Sur la somme directe $V\oplus V^*$, il existe une forme bilinéaire alternée $<.,.>$ et une seule telle que V et V^* soient totalement isotropes et que $<(x,0),(0,x')>=<x,x'>$ pour $x\in V, x'\in V^*$. Cette forme est non dégénérée. Nous noterons $A(V)$ l'algèbre de Weyl correspondante. Elle est engendrée comme algèbre par $V\cup V^*$. Elle contient les algèbres symétriques $S(V)$ et $S(V^*)$ comme sous-algèbres. Si $v\in V$ et $v'\in V^*$, on a, dans A(V),

$$[v,v']=<v,v'>.1.$$

Si $(p_1,\ldots,p_n)$ est une base de V et $(q_1,\ldots,q_n)$ la base duale de V^* , les p_i et les q_j forment un système générateur de $A(V)$ tel que

$$[p_i,p_j]=[q_i,q_j]=0 \qquad [p_i,q_j]=\delta_{ij}$$

quels que soient i et j . L'algèbre $A(V)$ s'identifie à l'algèbre des opérateurs différentiels à coefficients polynomiaux sur V . Dans cette identification, un élément v de V correspond à la dérivation définie par v , et un élément v' de V^* correspond à l'opérateur de multiplication par la fonction v' . Un élément de $A(V)$ qui appartient au sous-espace vectoriel $S(V^*).V$ correspond à un champ de vecteurs polynomial sur V .

1.2. Lemme. Soient W un sous-espace vectoriel de V, $V'=V/W$, $W°$ l'orthogonal de W dans V^* ; on identifie V'^* à $W°$. Soit B la sous-algèbre de $A(V)$ engendrée par V et $W°$.

(i) W est central dans B . Soit I l'idéal bilatère de B engendré par W .

(ii) Il existe un homorphisme φ et un seul de B dans $A(V')$ tel que

$$\varphi(x)=x \text{ mod.} W \qquad \text{pour tout } x\in V$$
$$\varphi(y)=y \qquad \text{pour tout } y\in W° .$$

Cet homorphisme est surjectif de noyau I , de sorte que l'algèbre B/I s'identifie canoniquement à $A(V')$.

Soient $(p_1,\ldots,p_n),(q_1,\ldots,q_n)$ comme en 1.1. On suppose que $(p_{m+1},\ldots,p_n)$ est une base de W . Alors $(q_1,\ldots,q_m)$ est une base de $W°$. L'algèbre B est engendrée par $p_1,\ldots,p_n,q_1,\ldots,q_m$, et (i) est évident. Pour tout $x\in V$, soit x' son image canonique dans V' . Tout $b\in B$ s'écrit de manière unique

$$\sum_{\alpha\beta}\lambda_{\alpha\beta}p_1^{\alpha_1}\ldots p_n^{\alpha_n}q_1^{\beta_1}\ldots q_m^{\beta_m}$$

où les $\lambda_{\alpha\beta}$ sont des scalaires. Notons $\varphi(b)$ l'élément

$$\sum_{\alpha\beta}\lambda_{\alpha\beta}p'^{\alpha_1}_1\ldots p'^{\alpha_n}_n q_1^{\beta_1}\ldots q_m^{\beta_m}$$
$$=\sum_{\alpha_{m+1}=\ldots=\alpha_n=0}\lambda_{\alpha\beta}p'^{\alpha_1}_1\ldots p'^{\alpha_m}_m q_1^{\beta_1}\ldots q_m^{\beta_m}$$

de $A(V')$. On vérifie facilement que φ est un homomorphisme d'algèbres, que φ

est surjectif, et que $\operatorname{Ker}\varphi = I$. L'assertion d'unicité du lemme est claire.

1.3. *Lemme. Soient* W *un sous-espace vectoriel de* V, W° *l'orthogonal de* W *dans* $V^\star$; *on identifie* $W^\star$ *à* $V^\star/W^\circ$. *Soit* C *la sous-algèbre de* $A(V)$ *engendrée par* W *et* $V^\star$.

(i) W° *est central dans* C . *Soit* J *l'idéal bilatère de* C *engendré par* W° .

(ii) *Il existe un homomorphisme* ψ *et un seul de* C *dans* $A(W)$ *tel que*

$$\psi(x)=x \quad \text{pour tout } x\in W$$
$$\psi(y)=y \bmod. W^\circ \quad \text{pour tout } y\in V^\star .$$

Cet homomorphisme est surjectif de noyau J , *de sorte que l'algèbre* C/J *s'identifie canoniquement à* $A(W)$.

Ce lemme se démontre comme en 1.2, et d'ailleurs se ramène si l'on veut à 1.2 (cf. 1.6).

1.4. Dans les conditions de 1.2 et 1.3, nous dirons que B (resp. C) est *la première (resp. deuxième) sous-algèbre de* $A(V)$ *définie canoniquement par* W , et que φ (resp. ψ) est *l'homomorphisme canonique de* B *sur* $A(V/W)$ (resp. *de* C *sur* $A(W)$)

1.5. Il existe un antiautomorphisme $a \mapsto a^\perp$ de $A(V)$ et un seul tel que $v^\perp = v$ pour tout $v\in V$ et $(v')^\perp = -v'$ pour tout $v\in V^\star$. De même, il existe un antiautomorphisme $a \mapsto a^\tau$ de $A(V)$ et un seul tel que $v^\tau = -v$ pour tout $v\in V$ et $(v')^\tau = v'$ pour tout $v'\in V^\star$. Si $a\in S(V^\star)V$ est un champ de vecteurs, on a $a^\tau = -a - \operatorname{div} a$. En effet, soient $p_1,\dots,q_n$ comme en 1.1. Si $a = q_1^{\alpha_1}\dots q_n^{\alpha_n} p_i$, on a $a^\tau = -p_i q_1^{\alpha_1}\dots q_n^{\alpha_n}$, donc $a + a^\tau = -[p_i, q_1^{\alpha_1}\dots q_n^{\alpha_n}] = -\frac{\partial}{\partial q_i}(q_1^{\alpha_1}\dots q_n^{\alpha_n})$.

1.6. Il existe un isomorphisme et un seul de $A(V)$ sur $A(V^\star)$ qui transforme v en v pour tout $v\in V$ et v' en $-v'$ pour tout $v'\in V^\star$. D'où un isomorphisme de $A(V)$ sur l'algèbre des opérateurs différentiels à coefficients polynomiaux sur $V^\star$. Dans cet isomorphisme, un élément v de V correspond à l'opérateur de multiplication par la fonction v sur $V^\star$, et un élément v' de $V^\star$ correspond à la dérivation définie par $-v'$.

1.7. Soient $(p_1,\dots,p_n)$ une base de V, $(q_1,\dots,q_n)$ la base duale de $V^\star$. Soient φ un endomorphisme de V , (α_{ij}) sa matrice par rapport à $(p_1,\dots,p_n)$. Alors φ définit le champ de vecteurs $x \mapsto \varphi(x)$ sur V , dont la valeur en $\lambda_1 p_1 + \dots + \lambda_n p_n$ est $\lambda_1\varphi(p_1) + \dots + \lambda_n\varphi(p_n)$. Ce champ de vecteurs, considéré comme élément de $A(V)$, est donc égal à $q_1\varphi(p_1) + \dots + q_n\varphi(p_n) = \sum_{ij}\alpha_{ij}q_j p_i$. Sa divergence est $\sum_i \alpha_{ii} = \operatorname{tr}\varphi$.

De même, ${}^t\varphi$ définit un champ de vecteurs dans $V^\star$, donc un élément de $A(V^\star)$ égal à $p_1\,{}^t\varphi(q_1) + \dots + p_n\,{}^t\varphi(q_n) = \sum_{ij}\alpha_{ji}p_j q_i = \sum_{ij}\alpha_{ij}p_i q_j = \sum_{ij}\alpha_{ij}q_j p_i + \sum_i \alpha_{ii}$. On voit donc que, *si* $\operatorname{tr}\varphi = 0$, *les champs définis par* φ *et* $-{}^t\varphi$ *se correspondent dans l'iso-*

morphisme de 1.6.

2. Champs invariants à droite ou à gauche sur une algèbre de Lie nilpotente.

Dans les §§2 et 3, $\underline{g}$ désigne une algèbre de Lie nilpotente sur k, et Γ le groupe adjoint de $\underline{g}$.

2.1. Nous allons utiliser l'algèbre $A(\underline{g})$. On a $\underline{g} \subset A(\underline{g})$, mais on prendra garde que, si $x, y \in \underline{g}$, le crochet de x, y dans $\underline{g}$ est distinct de leur crochet dans $A(\underline{g})$ (ce dernier est nul!). Aussi noterons-nous $\underline{[}x,y\underline{]}$ le crochet dans $\underline{g}$.

2.2. Soit G le groupe nilpotent qui admet $\underline{g}$ comme ensemble sous-jacent et la loi de Hausdorff comme loi de groupe. Si $x \in \underline{g}$, x définit un champ de vecteurs invariant à gauche $L_{\underline{g}}(x)$ sur G. Il est bien connu (du moins pour $k=\mathbb{R}$) que la valeur de $L_{\underline{g}}(x)$ en $y \in \underline{g}$ est

$$\frac{\mathrm{ad}\, y}{1-\exp(-\mathrm{ad}\, y)} x = x + \frac{1}{2}\underline{[}y,x\underline{]} + \ldots$$
$$= x + \sum_{r \geq 1} b_r (\mathrm{ad}\, y)^r x$$

($b_1 = \frac{1}{2}, b_2 = \frac{1}{12}, b_3 = b_5 = b_7 = \ldots = 0$); la série n'a qu'un nombre fini de termes non nuls. Le champ $L_{\underline{g}}(x)$ est polynomial, de sorte que $L_{\underline{g}}(x) \in S(\underline{g}^*)\underline{g} \subset A(\underline{g})$. L'application $L_{\underline{g}}$ est un homomorphisme de $\underline{g}$ dans l'algèbre de Lie $A(\underline{g})$. Cet homomorphisme se prolonge en un homomorphisme, que nous noterons encore $L_{\underline{g}}$, ou simplement L, de $U(\underline{g})$ dans l'algèbre associative $A(\underline{g})$.

Soient $(p_1, \ldots, p_n)$ une base de $\underline{g}$, $(q_1, \ldots, q_n)$ la base duale de $\underline{g}^*$. Si $y = \lambda_1 p_1 + \ldots + \lambda_n p_n \in \underline{g}$, on a

$$(\mathrm{ad}\, y)^r x = \sum_{1 \leq i_1, \ldots, i_r \leq n} \lambda_{i_1} \ldots \lambda_{i_r} (\mathrm{ad}\, p_{i_1}) \ldots (\mathrm{ad}\, p_{i_r}) x$$

donc

$$L_{\underline{g}}(x) = x + \sum_{r \geq 1} b_r \sum_{i_1, \ldots, i_r} q_{i_1} \ldots q_{i_r} \underline{[}\, p_{i_1}, \ldots, \underline{[}\, p_{i_r}, x\underline{]} \ldots \underline{]}.$$

2.3. Si $x \in \underline{g}$, x définit un champ de vecteurs invariant à droite $R_{\underline{g}}(x)$ sur G. La valeur de $R_{\underline{g}}(x)$ en $y \in \underline{g}$ est

$$\frac{\mathrm{ad}\, y}{\exp(\mathrm{ad}\, y) - 1} x = x - \frac{1}{2}\underline{[}y,x\underline{]} + \ldots$$
$$= x + \sum_{r \geq 1} b'_r (\mathrm{ad}\, y)^r x$$

($b'_1 = -\frac{1}{2}$, $b'_r = b_r$ pour $r \geq 2$). L'application $R_{\underline{g}}$ définit par prolongement un antihomomorphisme, que nous noterons encore $R_{\underline{g}}$, ou simplement R, de $U(\underline{g})$ dans $A(\underline{g})$.

Les sous-algèbres $L_{\underline{g}}(U(\underline{g}))$ et $R_{\underline{g}}(U(\underline{g}))$ de $A(\underline{g})$ sont permutables.

2.4. Pour tout $x\in\underline{g}$, nous poserons

$$L_{\underline{g}}(x)-R_{\underline{g}}(x)=W_{\underline{g}}(x)=W(x) .$$

La valeur de $W_{\underline{g}}(x)$ en $y\in\underline{g}$ est $[y,x]$. L'application $x\mapsto W_{\underline{g}}(x)$ de $\underline{g}$ dans $A(\underline{g})$ est un homomorphisme d'algèbres de Lie. Avec les notations de 2.2, on a

$$W_{\underline{g}}(x)=q_1\,[p_1,x]+\ldots+q_n\,[p_n,x] .$$

2.5. le groupe Γ opère naturellement dans $\underline{g},\underline{g}^*$, $S(\underline{g})$, $S(\underline{g}^*)$, $A(\underline{g})$, $U(\underline{g})$, d'où des représentations de Γ par automorphismes. Les applications $L_{\underline{g}}$, $R_{\underline{g}}$ de $U(\underline{g})$ dans $A(\underline{g})$ sont des homomorphismes de Γ-modules.

L'action correspondante de $\underline{g}$ dans $\underline{g},\underline{g}^*,\ldots,U(\underline{g})$ est l'action adjointe, d'où des représentations de $\underline{g}$ par dérivations. Les applications $L_{\underline{g}}$, $R_{\underline{g}}$ de $U(\underline{g})$ dans $A(\underline{g})$ sont des homomorphismes de $\underline{g}$-modules.

<u>Lemme</u>. <u>Si</u> $a\in A(\underline{g})$ <u>et</u> $x\in\underline{g}$, <u>le transformé</u> $x.a$ <u>de</u> a <u>par l'action adjointe de</u> x <u>est</u> $[W(x),a]$.

Il suffit de le vérifier quand $a\in\underline{g}$ et quand $a\in\underline{g}^*$. Soient $(p_1,\ldots,p_n),(q_1,\ldots,q_n)$ comme en 2.2. Si $a\in\underline{g}$, on a

$$[W(x),a]=[\textstyle\sum_i q_i\,[p_i,x]\,,a]=\sum_i[q_i,a]\,[p_i,x]$$

$$=-\textstyle\sum_i\langle a,q_i\rangle[p_i,x]=-[\sum\langle a,q_i\rangle p_i,x]=[x,a]=x.a .$$

Si $a\in\underline{g}^*$, on a

$$[W(x),a]=[\textstyle\sum_i q_i\,[p_i,x]\,,a]=\sum q_i\langle[p_i,x],a\rangle\in\underline{g}^*$$

donc, pour tout $y\in\underline{g}$,

$$\langle y,[W(x),a]\rangle=\textstyle\sum\langle y,q_i\rangle\langle[p_i,x],a\rangle$$

$$=\langle[\textstyle\sum\langle y,q_i\rangle p_i,x]\,,a\rangle=\langle-[x,y]\,,a\rangle=\langle y,x.a\rangle$$

d'où $[W(x),a]=x.a$.

2.6. Reprenons les notations de 2.2. La forme différentielle $dq_1\ldots dq_n$ sur $\underline{g}=G$ est invariante par translations à droite et à gauche sur G. Donc, pour tout $x\in\underline{g}$, les champs de vecteurs $L_{\underline{g}}(x)$, $R_{\underline{g}}(x)$ sont de divergence nulle. Par suite,

$$L_{\underline{g}}(x)=x+\textstyle\sum_{r\geq 1}b_r\sum_{i_1,\ldots,i_r}[p_{i_1},\ldots,[p_{i_r},x]\ldots]\,q_{i_1}\ldots q_{i_r}$$

$$W_{\underline{g}}(x)=[p_1,x]\,q_1+\ldots+[p_n,x]\,q_n .$$

2.7.<u>Lemme</u>. <u>Pour tout</u> $u\in U(\underline{g})$, <u>on a</u> $L(u)^{\perp}=R(u)$ <u>(cf.1.5)</u>.

Il suffit de le vérifier quand $u=x\in\underline{g}$. Alors, avec les notations de 2.2, on a

$$L(x)^{\perp}=(x+\tfrac{1}{2}\textstyle\sum_i q_i\,[p_i,x]+\sum_{r\geq 1}b_{2r}\sum_{i_1,\ldots,i_{2r}}q_{i_1}\ldots q_{i_{2r}}\,[p_{i_1},\ldots,[p_{i_{2r}},x]\ldots])^{\perp}$$

$$=x-\frac{1}{2}\sum_i [p_i,x] q_i+\sum_{r\geq 1} b_{2r}\sum_{i_1,\ldots,i_{2r}} [p_{i_1},\ldots,[p_{i_{2r}},x]\ldots]q_{i_1}\ldots q_{i_{2r}}$$

et cela est égal à $R(x)$ d'après 2.6.

2.8. Rappelons qu'il existe un antiautomorphisme $u\mapsto u^{\tau}$ de $U(\underline{g})$ et un seul tel que $x^{\tau}=-x$ pour tout $x\in\underline{g}$.

Lemme. Pour tout $u\in U(\underline{g})$, on a $L(u^{\tau})=L(u)^{\tau}$, $R(u^{\tau})=R(u)^{\tau}$ (cf.1.5).

Il suffit de le vérifier quand $u=x\in\underline{g}$. Alors $L(x^{\tau})=L(-x)=-L(x)=L(x)^{\tau}$ d'après 2.6, et de même pour R.

3. Passage d'un idéal de $S(\underline{g})$ à un idéal bilatère de $U(\underline{g})$.

3.1. Nous noterons $\Lambda(\underline{g})$ (resp. $P(\underline{g})$) l'idéal à gauche (resp. à droite) de $A(\underline{g})$ engendré par les $W(x)$ quand x parcourt $\underline{g}$. D'après 2.5, $\Lambda(\underline{g})$ et $P(\underline{g})$ sont invariants par Γ.

3.2. Lemme. Pour tout $u\in U(\underline{g})$, on a $R(u)-L(u)\in\Lambda(\underline{g})\cap P(\underline{g})$.

C'est clair si u est scalaire. Raisonnons par récurrence sur la filtration de u, Soit $u=xv$ avec $x\in\underline{g}$, $v\in U(\underline{g})$ et supposons le lemme vrai pour v. Alors, modulo $\Lambda(\underline{g})$, on a

$$\begin{aligned}R(xv)&=R(v)R(x)=R(v)(L(x)-W(x))\equiv R(v)L(x)\\&=L(x)R(v)=L(x)L(v)=L(xv).\end{aligned}$$

On raisonne de manière analogue pour $P(\underline{g})$.

3.3. Proposition. Soit J un idéal de $S(\underline{g})$.

(i) On a

$$\begin{aligned}L^{-1}(A(\underline{g})J+\Lambda(\underline{g}))&=R^{-1}(A(\underline{g})J+\Lambda(\underline{g}))\\&=L^{-1}(JA(\underline{g})+P(\underline{g}))=R^{-1}(JA(\underline{g})+P(\underline{g})).\end{aligned}$$

Cet ensemble est un idéal bilatère de $U(\underline{g})$ qu'on notera $\mu(J)$.

(ii) $\mu(J^{\tau})=(\mu(J))^{\tau}$.

(iii) Soit $J'=\sum_{\gamma\in\Gamma}\gamma J$. On a $A(\underline{g})J+\Lambda(\underline{g})=A(\underline{g})J'+\Lambda(\underline{g})$, et $\mu(J)=\mu(J')$.

La première égalité de (i) résulte de 3.2. D'autre part, $P(\underline{g})=\Lambda(\underline{g})^{\perp}$ d'après 2.6, donc

$$L^{-1}(JA(\underline{g})+P(\underline{g}))=L^{-1}((A(\underline{g})J+\Lambda(\underline{g}))^{\perp})=R^{-1}(A(\underline{g})J+\Lambda(\underline{g}))$$

d'après 2.7. De même,

$$R^{-1}(JA(\underline{g})+P(\underline{g})))=L^{-1}(A(\underline{g})J+\Lambda(\underline{g})).$$

Cela prouve (i).

On a $P(\underline{g})=\Lambda(\underline{g})^{\tau}$ d'après 2.6, d'où

$$\begin{aligned}\mu(J^{\tau})&=L^{-1}(A(\underline{g})J^{\tau}+\Lambda(\underline{g}))=L^{-1}((JA(\underline{g})+P(\underline{g}))^{\tau})\\&=(L^{-1}(JA(\underline{g})+P(\underline{g})))^{\tau}\quad\text{d'après 2.8}\end{aligned}$$

$=(\mu(J))^{\tau}$.

Enfin, si $x\in \underline{g}$ et $a\in A(\underline{g})J+\Lambda(\underline{g})$, on a

$$W(x)a-aW(x)\in W(x)(A(\underline{g})J+\Lambda(\underline{g}))+aW(\underline{g})$$
$$\subset A(\underline{g})J+\Lambda(\underline{g}) ;$$

d'après 2.5, il en résulte que $A(\underline{g})J+\Lambda(\underline{g})$ est Γ-invariant . Par suite, $\gamma J\subset A(\underline{g})J+\Lambda(\underline{g})$ pour tout $\gamma\in\Gamma$, d'où $A(\underline{g})J'+\Lambda(\underline{g})\subset A(\underline{g})J+\Lambda(\underline{g})$. Cela prouve (iii)

3.4. En vertu de 3.3(iii) , il suffit d'envisager les idéaux <u>Γ-invariants</u> de $S(\underline{g})$.

3.5.<u>Remarque</u>. Soit I un idéal bilatère de $U(\underline{g})$. On a

$$(A(\underline{g})L(I)+\Lambda(\underline{g}))\cap S(\underline{g})=(A(\underline{g})R(I)+\Lambda(\underline{g}))\cap S(\underline{g})$$
$$=(L(I)A(\underline{g})+P(\underline{g}))\cap S(\underline{g})=(R(I)A(\underline{g})+P(\underline{g}))\cap S(\underline{g}) .$$

Cet ensemble est un idéal Γ-invariant $\nu(I)$ de $S(\underline{g})$, et $\nu(I^{\tau})=(\nu(I))^{\tau}$.

Cela se démontre comme 3.3, en utilisant 2.5, 2.6, 2.7, 2.8.

Il serait souhaitable de prouver que, pour k algébriquement clos, μ et ν définissent des bijections réciproques entre l'ensemble des idéaux premiers de $U(\underline{g})$ et l'ensemble des idéaux premiers Γ-invariants de $S(\underline{g})$. Pour un résultat très partiel dans cette direction, cf.5.6.

4.<u>Un lemme sur les opérateurs différentiels à coefficients polynomiaux.</u>

4.1. Soit G un groupe opérant dans un espace vectoriel complexe de dimension n par des applications biholomorphes de V dans V . On dira que G opère de manière unimodulaire dans V si G conserve la forme différentielle $dz_1dz_2\dots dz_n$ sur V (où$(z_1,\dots,z_n)$ désigne une base de V^{*} ; cette définition est indépendante du choix de la base). S'il en est ainsi, et si $V_{\mathbb{R}}$ désigne l'espace vectoriel réel sous-jacent à V, G conserve la mesure de Lebesgue de $V_{\mathbb{R}}$.

4.2. Soit X un espace vectoriel réel de dimension finie. Si a est un opérateur différentiel sur X à coefficients C^{∞} , et si T est une distribution sur X, a.T est une distribution bien définie sur X . Soit b un champ de vecteurs à coefficients C^{∞} sur X ; il y a deux manières de définir b.T : On peut considérer b comme un opérateur différentiel, d'où une action de b sur T ; et l'on peut considérer la dérivée de Lie de T pour l'action du groupe à un paramètre défini par b ; toutefois les deux sens possibles de b.T coïncident si ce groupe à un paramètre conserve la mesure de Lebesgue de X .

4.3.<u>Lemme</u>. <u>Soient</u> V <u>un espace vectoriel complexe de dimension finie</u>, G <u>un groupe algébrique complexe opérant dans</u> V <u>de manière unimodulaire et régulièrement au sens de la géométrie algébrique</u>. <u>Soit</u> $\underline{g}$ <u>l'algèbre de Lie de</u> G . <u>Chaque</u> $x\in\underline{g}$ <u>dé-</u>

finit un champ de vecteurs à coefficients polynomiaux sur V . Soit Δ l'idéal à gauche de A(V) engendré par ces champs.

Soit W une sous-variété fermée au sens de Zariski de V , lisse, et G- stable. Soit Ξ l'ensemble des G-orbites contenues dans W . On suppose que ces orbites sont toutes de même dimension (donc fermées au sens de Zariski), et que tout $\Omega \in \Xi$ porte une mesure positive μ_Ω non nulle et G-invariante. On considère ces μ_Ω comme des distributions sur $V_{\mathbb{R}}$.

Pour tout sous-ensemble E de V , soit J(E) l'ensemble des éléments de $S(V^*)$ qui s'annulent sur E .

Soit $a \in A(V)$. Les conditions suivantes sont équivalentes :

(i) $a.\mu_\Omega = 0$ pour tout $\Omega \in \Xi$;

(ii) $a \in A(V)J(\Omega)+\Delta$ pour tout $\Omega \in \Xi$;

(iii) $a \in A(V)J(W)+\Delta$.

(Je remercie J.L.Verdier pour son aide dans la démonstration de ce lemme).

(iii) $\Longrightarrow$(ii). C'est évident.

(ii) $\Longrightarrow$(i). Soit $\Omega \in \Xi$. Si $a \in J(\Omega)$, il est clair que $a.\mu_\Omega = 0$.
Si a est le champ de vecteurs défini par un $x \in \underline{g}$, les deux sens possibles de $a.\mu_\Omega$ coïncident puisque G opère de manière unimodulaire; comme G laisse invariante μ_Ω , on a encore $a.\mu_\Omega = 0$.

(i) $\Longrightarrow$ (iii). Supposons $a.\mu_\Omega = 0$ pour tout $\Omega \in \Xi$ et prouvons que $a \in A(V)J(W)+\Delta$. Si a est de degré 0 , c'est-à-dire est la multiplication par une fonction polynomiale sur V , il est clair que $a \in J(W) \subset A(V)J(W)+\Delta$. Supposons notre assertion établie quand a est de degré $<p$, et envisageons le cas où a est de degré $p>0$.

Soit T(V) le fibré vectoriel de base V dont la fibre en v est l'ensemble des distributions sur V concentrées en v et d'ordre $\leq p$ (au sens de [1],13.2 ; ici, on ne considère pas V comme variété réelle mais comme variété complexe). Les sections de T(V) régulières au sens de la géométrie algébrique s'identifient aux opérateurs différentiels réguliers sur V de degré $\leq p$. On définit de même T(W) , et $T(\Omega)$ pour tout $\Omega \in \Xi$. Pour tout $v \in W$, la fibre $T(W)_v$ s'identifie à un sous-espace vectoriel de $T(V)_v$; pour tout $\Omega \in \Xi$ et tout $v \in \Omega$, $T(\Omega)_v$ s'identifie à un sous-espace vectoriel de $T(W)_v$.

Soit $\Omega \in \Xi$. On a $a.\mu_\Omega = 0$. Par suite, pour tout $v \in \Omega$, l'élément de $T(V)_v$ défini par a appartient à $T(\Omega)_v$ (cela résulte de [9],p.102).

Soit $(x_1,\ldots,x_t)$ une base de $\underline{g}$. Soient $c_1,\ldots,c_t$ les champs de vecteurs correspondants sur V .

Soit $\mathcal{C}(W)$ le faisceau des germes de sections régulières de T(W) . Soit $\mathcal{C}'(W)$ le faisceau des germes de sections régulières s de T(W) telles que $s(v) \in T(\Omega)_v$ pour tout $\Omega \in \Xi$ et tout $v \in \Omega$ où s est défini. Soit $\mathcal{O}(W)$ le faisceau des germes de fonctions régulières sur W . Pour $\alpha_1,\ldots,\alpha_t \in \mathbb{N}$ et $\alpha_1+\ldots+\alpha_t \leq p$, soit $f_{\alpha_1\ldots\alpha_t}$ une fonction régulière sur un ouvert de Zariski W' de W . Alors l'opérateur différentiel $\sum_{\alpha_1,\ldots,\alpha_t} f_{\alpha_1\ldots\alpha_t} c_1^{\alpha_1}\ldots c_t^{\alpha_t} | W'$ est une section régulière de $\mathcal{E}'(W)$ sur un ouvert de Zariski de W. D'où un homomorphisme

(1) $$\mathcal{O}(W)^{\binom{p+t}{t}} \rightarrow \mathcal{E}'(W) .$$

Comme, pour tout $v \in \Omega \in \Xi$, les valeurs de $c_1,\ldots,c_t$ en v engendrent l'espace tangent à Ω, cet homomorphisme est surjectif sur les fibres. Les faisceaux $\mathcal{O}(W)$ et $\mathcal{E}'(W)$ sont cohérents. Soit $\mathcal{J}$ le noyau de l'homomorphisme (1). Comme $H^1(W, \mathcal{J})=0$ (car W est affine), l'homomorphisme

$$H^\circ(W, \mathcal{O}(W)^{\binom{p+t}{t}}) \longrightarrow H^\circ(W, \mathcal{E}'(W))$$

est surjectif. Il existe donc des fonctions $f_{\alpha_1\ldots\alpha_t}$ régulières sur W telles que

$$a|W = \sum_{\alpha_1,\ldots,\alpha_t} f_{\alpha_1\ldots\alpha_t} c_1^{\alpha_1}\ldots c_t^{\alpha_t}|W .$$

Autrement dit, il existe des $g_{\alpha_1\ldots\alpha_t}$ régulières sur V telles que $a-\sum_{\alpha_1,\ldots,\alpha_t} g_{\alpha_1\ldots\alpha_t} c_1^{\alpha_1}\ldots c_t^{\alpha_t}$ s'annule sur W. Soit $(p_1,\ldots,p_m)$ une base de V. On a alors

(2) $$a-\sum_{\alpha_1,\ldots,\alpha_t} g_{\alpha_1\ldots\alpha_t} c_1^{\alpha_1}\ldots c_t^{\alpha_t} = \sum_{\beta_1,\ldots,\beta_m \in \mathbb{N},\, \beta_1+\ldots+\beta_m \leq p} h_{\beta_1\ldots\beta_m} p_1^{\beta_1}\ldots p_m^{\beta_m}$$

où les $h_{\beta_1\ldots\beta_m}$ appartiennent à $J(W)$ (l'égalité (2) est une égalité dans $A(V)$). Posons

$$a' = \sum_{\alpha_1+\ldots+\alpha_t>0} g_{\alpha_1\ldots\alpha_t} c_1^{\alpha_1}\ldots c_t^{\alpha_t} + \sum_{\beta_1+\ldots+\beta_m \leq p} p_1^{\beta_1}\ldots p_m^{\beta_m} h_{\beta_1\ldots\beta_m} .$$

Il résulte de (2) que $a=a'+b$, où b est un opérateur différentiel régulier sur V de degré $\leq p-1$. Par ailleurs, $a' \in \Delta + A(V)J(W)$, donc $a'.\mu_\Omega=0$ pour tout $\Omega \in \Xi$ puisque (iii)$\Longrightarrow$(i). On en déduit que $b.\mu_\Omega=0$ pour tout $\Omega \in \Xi$, et il suffit d'appliquer à b l'hypothèse de récurrence.

4.4. Bien que ce ne soit pas utile pour la suite, montrons que l'hypothèse faite dans 4.3 sur les dimensions des orbites est indispensable.

Soit $\underline{g}$ l'algèbre de Lie complexe de dimension 6, admettant la base $(e_1,\ldots,e_6)$ et la table de multiplication

$$[e_1,e_2]=e_5 , \ [e_1,e_3]=e_6 , \ [e_2,e_4]=e_6 ,$$

les autres crochets étant nuls ou déduits des précédents par antisymétrie. Le centre de $\underline{g}$ est égal à $[\underline{g},\underline{g}]$ et à $\mathbb{C}e_5+\mathbb{C}e_6$. L'algèbre $\underline{g}$ est nilpotente. Soit G le groupe de Lie simplement connexe d'algèbre de Lie $\underline{g}$.

Avec les notations de 4.3, prenons d'abord $V=W=\underline{g}$, d'où $J(W)=0$, et faisons opérer G dans $\underline{g}$ par la représentation adjointe. Cette action est unimodulaire et régulière au sens de la géométrie algébrique. Les orbites sont fermées. Soient $x_1,\ldots,x_6$ les fonctions coordonnées sur $\underline{g}$ correspondant à $(e_1,\ldots,e_6)$. Les champs de vecteurs définis par $e_1,\ldots,e_6$ sont

$$-x_2 \frac{\partial}{\partial x_5} - x_3 \frac{\partial}{\partial x_6} , x_1 \frac{\partial}{\partial x_5} - x_4 \frac{\partial}{\partial x_6} , x_1 \frac{\partial}{\partial x_6} , x_2 \frac{\partial}{\partial x_6} , 0 , 0 .$$

Les orbites Ω dans g sont de dimensions 2,1,0, faciles à calculer ainsi que leurs mesures invariantes μ_Ω. On vérifie sans peine que le champ de vecteurs $x_1 \frac{\partial}{\partial x_5}$ sur g est en chaque point de g tangent à l'orbite de ce point et annule toutes les μ_Ω. Pourtant, $x_1 \frac{\partial}{\partial x_5} \notin \Delta$ (voir plus bas).

Prenons maintenant $V=W=g^*$ et faisons opérer G dans g^* par la représentation coadjointe. Cette action est unimodulaire et régulière au sens de la géométrie algébrique. Les orbites sont fermées. Soient $y_1,\ldots,y_6$ les fonctions coordonnées sur g^* correspondant à la base duale de $(e_1,\ldots,e_6)$. Les champs de vecteurs définis par $e_1,\ldots,e_6$ sont

$$y_5 \frac{\partial}{\partial y_2} + y_6 \frac{\partial}{\partial y_3} , -y_5 \frac{\partial}{\partial y_1} + y_6 \frac{\partial}{\partial y_4} , -y_6 \frac{\partial}{\partial y_1} , -y_6 \frac{\partial}{\partial y_2} , 0 , 0 .$$

Les orbites Ω dans g^* sont de dimension 4,2,0, faciles à calculer ainsi que leurs mesures de Kostant μ_Ω. On vérifie que le champ $y_5 \frac{\partial}{\partial y_1}$ sur g^* est en chaque point tangent à l'orbite de ce point et annule les μ_Ω. Mais on va voir que $y_5 \frac{\partial}{\partial y_1} \notin \Delta$, (d'où en même temps le résultat négatif annoncé ci-dessus).

Il suffit de prouver notre assertion en remplaçant partout le corps $\mathbb{C}$ par le corps $\mathbb{R}$. Soit T la distribution $y_4 y_5 \frac{\partial}{\partial y_6} \delta(y_6) - y_1 \delta(y_6)$ sur g^* (on note δ la fonction de Dirac à l'origine). On a $y_6 T = -y_4 y_5 \delta(y_6)$, d'où

$$0 = -y_6 \frac{\partial}{\partial y_1} T = -y_6 \frac{\partial}{\partial y_2} T = (y_5 \frac{\partial}{\partial y_2} + y_6 \frac{\partial}{\partial y_3}) T$$

$$y_5 \frac{\partial}{\partial y_1} T = -y_5 \delta(y_6)$$

$$y_6 \frac{\partial}{\partial y_4} T = \frac{\partial}{\partial y_4}(-y_4 y_5 \delta(y_6)) = -y_5 \delta(y_6) .$$

Ainsi, T est annulée par tous les éléments de Δ, mais pas par $y_5 \frac{\partial}{\partial y_1}$, d'où $y_5 \frac{\partial}{\partial y_1} \notin \Delta$.

4.5. Conservons les notations de 4.4. Comme $y_5 \frac{\partial}{\partial y_1}$ annule les mesures de Kostant, mais pas T, on voit que la distribution T, bien que G-invariante, n'est pas limite de combinaisons linéaires de mesures de Kostant. Un tel contre-exemple avait été construit en [3]. Mais ici l'impossibilité d'obtenir T comme limite de combinaisons linéaires de mesures de Kostant est vraie même localement, contrairement à ce qui se passe dans [3].

5. Le résultat principal.

5.1. Lemme. *Soient g une algèbre de Lie nilpotente sur* k , *a un idéal central de g , g'=g/a , B la première sous-algèbre de* A(g) *définie par a* (cf. 1.4), π *l'ho-*

momorphisme canonique de B sur $A(\underline{g}')$, et $u \mapsto u'$ l'homomorphisme canonique de $U(\underline{g})$ sur $U(\underline{g}')$.

(i) Pour tout $u \in U(\underline{g})$, on a $L_{\underline{g}}(u), R_{\underline{g}}(u) \in B$, et $\pi(L_{\underline{g}}(u)) = L_{\underline{g}'}(u')$, $\pi(R_{\underline{g}}(u)) = R_{\underline{g}'}(u')$.

(ii) Si J est un sous-espace vectoriel de $S(\underline{g})$, on a $(A(\underline{g})J + A(\underline{g})) \cap B = BJ + BW_{\underline{g}}(\underline{g})$.

Il suffit de prouver (i) quand $u = x \in \underline{g}$. Soit $(p_1, \ldots, p_n)$ une base de $\underline{g}$ telle que $(p_{m+1}, \ldots, p_n)$ soit une base de $\underline{a}$. Soit $(q_1, \ldots, q_n)$ la base duale de $\underline{g}^*$, de sorte que B est engendrée par $p_1, \ldots, p_n, q_1, \ldots, q_m$. Si l'un des entiers $i_1, \ldots, i_r$ est $>m$, on a $[p_{i_1}, \ldots, [p_{i_r}, x] \ldots] = 0$. Donc

$$L_{\underline{g}}(x) = x + \sum_{r \geq 1} b_r \sum_{1 \leq i_1, \ldots, i_r \leq m} q_{i_1} \ldots q_{i_r} [p_{i_1}, \ldots, [p_{i_r}, x] \ldots] \in B$$

et

$$\pi(L_{\underline{g}}(x)) = x' + \sum_{r \geq 1} b_r \sum_{1 \leq i_1, \ldots, i_r \leq m} q_{i_1} \ldots q_{i_r} [p'_{i_1}, \ldots, [p'_{i_r}, x'] \ldots] = L_{\underline{g}'}(x') .$$

On raisonne de même pour $R_{\underline{g}}(x)$.

Considérons $A(\underline{g})$ comme un B-module à droite. Alors $A(\underline{g})$ est libre et admet pour base les monômes en $q_{m+1}, \ldots, q_n$. Soit J un sous-espace vectoriel de $S(\underline{g})$. On a $J \subset S(\underline{g}) \subset B$, et $W_{\underline{g}}(\underline{g}) \subset B$ d'après (i) , donc

$$A(\underline{g})J + A(\underline{g}) = A(\underline{g})(BJ + BW_{\underline{g}}(\underline{g})) = \oplus_{\alpha} q_{m+1}^{\alpha_{m+1}} \ldots q_n^{\alpha_n} (BJ + BW_{\underline{g}}(\underline{g}))$$

d'où (ii) .

5.2. Lemme. Soient $\underline{g}$ une algèbre de Lie nilpotente sur k , $\underline{g}'$ un idéal de $\underline{g}$, B la deuxième sous-algèbre de $A(\underline{g})$ définie par $\underline{g}'$ (cf. 1.4), π l'homomorphisme canonique de B sur $A(\underline{g}')$. Pour tout $u \in U(\underline{g}') \subset U(\underline{g})$, on a $L_{\underline{g}}(u), R_{\underline{g}}(u) \in B$, et $\pi(L_{\underline{g}}(u)) = L_{\underline{g}'}(u), \pi(R_{\underline{g}}(u)) = R_{\underline{g}'}(u)$.

Il suffit de le prouver quand $u = x \in \underline{g}'$. Soit $(p_1, \ldots, p_n)$ une base de $\underline{g}$ telle que $(p_{m+1}, \ldots, p_n)$ soit une base de $\underline{g}'$. Soit $(q_1, \ldots, q_n)$ la base duale de $\underline{g}^*$, de sorte que B est engendrée par $p_{m+1}, \ldots, p_n, q_1, \ldots q_n$. On a

$$L_{\underline{g}}(x) = x + \sum_{r \geq 1} b_r \sum_{1 \leq i_1, \ldots, i_r \leq n} q_{i_1} \ldots q_{i_r} [p_{i_1}, \ldots, [p_{i_r}, x] \ldots] \in B$$

car $[\underline{g}', \underline{g}] \subset \underline{g}'$. Le noyau de π est engendré par $q_1, \ldots, q_m$. Notant $f \mapsto f'$ le morphisme canonique de $\underline{g}^*$ sur $\underline{g}'^*$, on a

$$\pi(L_{\underline{g}}(x)) = x + \sum_{r \geq 1} b_r \sum_{m+1 \leq i_1, \ldots, i_r \leq n} q'_{i_1} \ldots q'_{i_r} [p_{i_1}, \ldots, [p_{i_r}, x] \ldots]$$
$$= L_{\underline{g}'}(x) .$$

On raisonne de même pour $R_{\underline{g}}(x)$.

5.3. Soit $\underline{g}$ une algèbre de Lie nilpotente sur k . Rappelons que, si ω est une orbite du groupe adjoint Γ dans $\underline{g}^*$, on sait construire canoniquement dans

$U(\underline{g})$ un idéal rationnel (c'est-à-dire primitif si k est algébriquement clos) $I(\omega)$, et que $\omega \mapsto I(\omega)$ est une bijection de $\underline{g}^*/\Gamma$ sur l'ensemble des idéaux rationnels de $U(\underline{g})$; cf. par exemple [2], chap.VI.

Théorème. Soient $\underline{g}$ une algèbre de Lie nilpotente sur k, Γ le groupe adjoint de $\underline{g}$, ω une Γ-orbite dans $\underline{g}^*$, $J(\omega)$ l'idéal de $S(\underline{g})$ correspondant à ω. Alors $\mu(J(\omega))=I(\omega)$ (cf.3.4).

a) Soit k' une extension de k. Par passage de k à k', on déduit de $\underline{g},\Gamma,\ldots$ des objets $\underline{g}',\Gamma',\ldots$. On a $I(\omega')=I(\omega)\otimes_k k'$, $J(\omega')=J(\omega)\otimes_k k'$, $\Lambda(\underline{g}')=\Lambda(\underline{g})\otimes_k k'$, $\mu(J(\omega'))=\mu(J(\omega))\otimes_k k'$. Le th.5.3 et l'énoncé analogue pour k' sont donc équivalents. Or il existe un sous-corps k_1 de type fini de k, une algèbre de Lie $\underline{g}_1$ sur k_1 et une orbite ω_1 dans $\underline{g}_1^*$ tels que $\underline{g},\omega$ se déduisent de $\underline{g}_1,\omega_1$ par extension des scalaires. On se ramène donc en deux étapes au cas où $k=\mathbb{C}$, ce que nous supposons désormais. A partir de là, l'usage de 4.3 et des méthodes de [7] permettrait sans doute de prouver 5.3. Nous procéderons autrement.

Le théorème est évident pour $\underline{g}$ commutatif. On raisonne par récurrence sur $\dim \underline{g} \geq 3$.

b) On suppose qu'un élément de ω, donc tous les éléments de ω, s'annulent sur un idéal central non nul $\underline{a}$ de $\underline{g}$. Alors $\underline{a}\subset I(\omega)$ et $\underline{a}\subset J(\omega)$.
Si $x\in\underline{a}$, on a $L(x)=x\in J(\omega)\subset A(\underline{g})J(\omega)+\Lambda(\underline{g})$, donc $x\in\mu(J(\omega))$; ainsi, $\underline{a}\subset\mu(J(\omega))$.

Soient $\underline{g}'=\underline{g}/\underline{a}$, B la première sous-algèbre de $A(\underline{g})$ définie par $\underline{a}$, π l'homomorphisme canonique de B sur $A(\underline{g}')$. Alors

(3) $\quad\quad \operatorname{Ker}\pi=B\underline{a}\subset BJ(\omega)$.

On a $\omega\subset\underline{g}'^*\subset\underline{g}^*$, $S(\underline{g})\subset B$, et $\pi(J(\omega))$ est l'idéal J' de $S(\underline{g}')$ défini par ω. D'autre part, $\pi(W_{\underline{g}}(\underline{g}))=W_{\underline{g}'}(\underline{g}')$ d'après 5.1(i).

Soient $u\in U(\underline{g})$, u' son image canonique dans $U(\underline{g}')$. On sait que $I(\omega)$ est l'image réciproque de $I(\omega)$ dans $U(\underline{g})$ (cf. par exemple [2],5.1.12). Alors

$$\begin{aligned} u\in\mu(J(\omega)) &\Longleftrightarrow L_{\underline{g}}(u)\in A(\underline{g})J(\omega)+\Lambda(\underline{g}) \\ &\Longleftrightarrow L_{\underline{g}}(u)\in BJ(\omega)+BW_{\underline{g}}(\underline{g}) \quad \text{d'après 5.1,(i) et (ii)} \\ &\Longleftrightarrow L_{\underline{g}'}(u')\in A(\underline{g}')\pi(J(\omega))+\Lambda(\underline{g}')\pi(W_{\underline{g}}(\underline{g})) \quad \text{d'après (3)} \\ &\Longleftrightarrow L_{\underline{g}'}(u')\in A(\underline{g}')J'+\Lambda(\underline{g}') \\ &\Longleftrightarrow u'\in I(\omega) \quad \text{(hypothèse de récurrence)} \\ &\Longleftrightarrow u\in I(\omega) \end{aligned}$$

d'où le théorème dans ce cas.

c) On suppose qu'on n'est pas dans le cas b). Alors le centre $\underline{z}$ de $\underline{g}$ est de dimension 1. Introduisons classiquement un idéal commutatif $\underline{y}$ de dimension 2 de $\underline{g}$ contenant $\underline{z}$, le commutant $\underline{g}'$ de $\underline{y}$ dans $\underline{g}$, qui est un idéal de codimension 1, et des éléments x,y,z de $\underline{g}$ tels que $x\notin\underline{g}'$, $y\in\underline{y}-\underline{z}$, $z\in\underline{z}-\{0\}$, $[x,y]=z$. Soit $f\in\omega$. On a $f(z)\neq 0$. Par un bon choix de x,y,z, on peut supposer que $f(y)=0$, $f(z)=1$.

Soient Γ' le groupe adjoint de $\underline{g}'$, $f'=f|\underline{g}'$, $\omega'=\Gamma'f'$, $\omega''=\bigcup_{\gamma\in\Gamma}\gamma\omega'=\Gamma f'\subset\underline{g}'^{*}$. Alors ω'' est une sous-variété fermée lisse de $\underline{g}'^{*}$, invariante par Γ . Pour tout $g\in\omega'$ et tout $\lambda\in k$, on a $(\Gamma'g)(y)=0$, et $((\exp \lambda x)g)(y)=g(y-\lambda z)=-\lambda$; donc l'application $(\lambda,g)\mapsto(\exp \lambda x)g$ de $k\times\omega'$ dans $\underline{g}'^{*}$ est un isomorphisme de la variété $k\times\omega'$ sur la variété ω'' . D'autre part, si $h\in\underline{g}^{*}$ est tel que $h(z)=1$, on a $((\exp \lambda y)h)(x)=h(x+\lambda z)=h(x)+\lambda$, donc ω est l'image réciproque de ω'' par l'application canonique $\underline{g}^{*}\to\underline{g}'^{*}$. Soient J',J'' les idéaux de $S(\underline{g}')$ correspondant à ω',ω'' . Alors $J''=\bigcap_{\gamma\in\Gamma}\gamma J'$, et $J(\omega)=S(\underline{g})J''$ d'après ce qui précède.

On sait que $I(\omega)$ est le plus grand idéal bilatère de $U(\underline{g})$ contenu dans $U(\underline{g})I(\omega')$ (cf. par exemple [2], démonstration de 6.1.7) . Soit $u\in I(\omega)$. Ecrivons $u=x^n u_n+x^{n-1}u_{n-1}+\dots+u_0$ avec $u_n,\dots,u_0\in U(\underline{g}')$. Alors $u_i\in I(\omega')$ pour tout i . Pour tout $\gamma\in\Gamma$, on a $\gamma u\in I(\omega)$, d'où $\gamma u_i\in I(\omega')$ pour tout i par récurrence sur n . Ainsi, $I(\omega)$ est l'idéal bilatère de $U(\underline{g})$ engendré par $\bigcap_{\gamma\in\Gamma}\gamma I(\omega')$. (Tout ce qu'on a dit jusqu'ici en c) est bien connu).

Appliquons 4.3 avec $V=\underline{g}^{*}$, $G=\Gamma$, $W=\omega''$. Identifions $A(\underline{g})$ à $A(\underline{g}^{*})=A(V)$ grâce à 1.6. Alors compte tenu de 1.7 , l'idéal à gauche noté Δ en 4.3 n'est autre que $\Lambda(\underline{g})$. Une partie facile du lemme 4.3 entraîne d'abord que $A(\underline{g})J(\omega)+\Lambda(\underline{g})\neq A(\underline{g})$, donc $\mu(J(\omega))\neq U(\underline{g})$. Si l'on prouve que $I(\omega)\subset\mu(J(\omega))$, la maximalité de $I(\omega)$ ([2],4.7.4) entraînera le théorème. Compte tenu de l'alinéa précédent, nous choisissons donc un élément u de $U(\underline{g}')$ tel que $\gamma u\in I(\omega')$ pour tout $\gamma\in\Gamma$, et il s'agit de montrer que $u\in\mu(J(\omega))$.

D'après l'hypothèse de récurrence, on a $L_{\underline{g}'}(\gamma u)\in A(\underline{g}')J'+\Lambda(\underline{g}')$ pour tout $\gamma\in\Gamma$, donc $L_{\underline{g}'}(u)\in A(\underline{g}')\gamma(J')+\Lambda(\underline{g}')$ pour tout $\gamma\in\Gamma$ (cf.3.1) . <u>D'après 4.3,</u> on en conclut que

(4) $$L_{\underline{g}'}(u)\in A(\underline{g}')J''+\Lambda(\underline{g}') .$$

Soit B la deuxième sous-algèbre de $A(\underline{g})$ définie par $\underline{g}'$. Soit π l'homomorphisme canonique de B sur $A(\underline{g}')$. D'après 5.2,(4) entraîne

(5) $$\pi(L_{\underline{g}}(u))\in\pi(BJ''+BW_{\underline{g}}(\underline{g}')) .$$

Soit $(p_1,\dots,p_n)$ une base de $\underline{g}$ telle que $p_1=x$, $p_2,\dots,p_n\in\underline{g}'$, $p_{n-1}=y$, $p_n=z$. Soit $(q_1,\dots,q_n)$ la base duale de $\underline{g}^{*}$. Alors B est engendrée par $p_2,\dots,p_n$, $q_1,\dots,q_n$, et l'idéal $\operatorname{Ker}\pi$ est engendré par q_1 . D'après 2.4, on a

$$W_{\underline{g}}(p_{n-1})=q_1[p_1,p_{n-1}]=q_1p_n .$$

Par ailleurs, $p_n=z$ prend la valeur constante 1 sur ω , donc $p_n-1\in J''$, donc

$$q_1=q_1p_n-q_1(p_n-1)\in W_{\underline{g}}(\underline{g}')+BJ''$$

d'où

$$\operatorname{Ker}\pi\subset BJ''+BW_{\underline{g}}(\underline{g}') .$$

Alors, (5) entraîne

$$L_{\underline{g}}(u)\in BJ''+BW_{\underline{g}}(\underline{g}')\subset A(\underline{g})J(\omega)+\Lambda(\underline{g})$$

d'où $u\in\mu(J(\omega))$.

5.4. En combinant 5.3 et 3.3(ii), on retrouve le résultat de [6], lemme 8.1, suivant lequel $I(-\omega)=I(\omega)^{\tau}$.

5.5. Supposons k algébriquement clos. Soit X une sous- variété fermée lisse de $\underline{g}^{*}$, invariante par Γ, et telle que toutes les Γ-orbites contenues dans X aient même dimension. Soit J l'idéal correspondant de $S(\underline{g})$. Soit I l'idéal bilatère de $U(\underline{g})$ associé à J (cf. par exemple [2], 6.4.8). Alors, 5.3 et 4.3 entraînent que $I=\mu(J)$. Il serait souhaitable de prouver cela sous la seule hypothèse que X est une sous-variété fermée de $\underline{g}^{*}$ invariante par Γ.

5.6. Remarque. Soit J un idéal premier Γ-invariant de $S(\underline{g})$. Alors :
(i) $(A(\underline{g})J+\Lambda(\underline{g}))\cap S(\underline{g})=J$; (ii) $\nu(\mu(J))\subset J$ avec les notations de 3.5.

En effet, comme en 5.3 a), on peut supposer $k=\mathbb{C}$. Soit X la sous-variété de $\underline{g}^{*}$ correspondant à J. Si $a\in A(\underline{g})J+\Lambda(\underline{g})$, et si l'on interprète a comme un opérateur différentiel sur $\underline{g}^{*}$, a annule les mesures Γ-invariantes des Γ-orbites contenues dans X. Si de plus $a\in S(\underline{g})$, l'opérateur différentiel a est la multiplication par une fonction; cette fonction doit alors s'annuler sur X, d'où $a\in J$ puisque J est premier. Ainsi, $(A(\underline{g})J+\Lambda(\underline{g}))\cap S(\underline{g})\subset J$. L'inclusion inverse est évidente, d'où (i). Comme $L(\mu(J))\subset A(\underline{g})J+\Lambda(\underline{g})$, on voit que $L(\mu(J))\cap S(\underline{g})\subset J$, d'où (ii).

Si J n'est pas premier, ou n'est pas Γ-invariant, (i) et (ii) peuvent être en défaut. Par exemple, prenons pour $\underline{g}$ l'algèbre de Heisenberg de base (p_1,p_2,p_3) avec $[p_1,p_2]=p_3$, $[p_1,p_3]=[p_2,p_3]=0$. Soit J l'idéal de $S(\underline{g})$ engendré par p_3^2 et p_1p_3. Il est Γ-invariant, mais pas premier. On a $p_3=-q_1(p_1p_3)+p_1(q_1p_3)\in$ $(A(\underline{g})J+\Lambda(\underline{g}))\cap S(\underline{g})$. D'autre part, $L(p_3)=p_3$, donc $p_3\in\mu(J)$ et par suite $p_3\in\nu(\mu(J))$. Mais $p_3\notin J$.

Notons que si J' est l'idéal de $S(\underline{g})$ engendré par p_3, on a $J\neq J'$ mais $\mu(J)=\mu(J')$.

6. Suggestions pour le cas non nilpotent.

6.1. Soit $\underline{g}$ une algèbre de Lie de dimension finie sur k. Soit $S\hat{}(\underline{g}^{*})$ l'algèbre des séries formelles sur $\underline{g}$, complétée de l'algèbre $S(\underline{g}^{*})$ des fonctions polynomiales. Soit $A\hat{}(\underline{g})$ l'algèbre des opérateurs différentiels sur $\underline{g}$ à coefficients séries formelles; elle est engendrée par les sous-algèbres $S(\underline{g})$ et $S\hat{}(\underline{g}^{*})$. Si $x\in\underline{g}$, le champ de vecteurs $L_{\underline{g}}(x)=L(x)$ défini par

$$(L_{\underline{g}}(x))(y)=\frac{\mathrm{ady}}{1-\exp(-\mathrm{ady})}\,x=x+\sum_{r\geq 1}b_r(\mathrm{ady})^r x$$

est cette fois un élément de $A\hat{}(\underline{g})$ (en principe, $(L_{\underline{g}}(x))(y)$ n'a pas de sens). L'application $L_{\underline{g}}$ est un homomorphisme de $\underline{g}$ dans l'algèbre de Lie $A\hat{}(\underline{g})$. Cet homomorphisme se prolonge en un homomorphisme noté encore $L_{\underline{g}}$, ou L, de $U(\underline{g})$

dans l'algèbre associative $A^{\wedge}(\underline{g})$. On définit de même $R_{\underline{g}}=R : U(\underline{g}) \to A^{\wedge}(\underline{g})$, et $W_{\underline{g}}(x)=W(x)=L(x)-R(x)\in A(\underline{g})$ pour tout $x\in \underline{g}$.

6.2. Nous allons écrire certaines identités dans $A^{\wedge}(\underline{g})$. Ces identités peuvent sans doute se prouver algébriquement. Mais le plus simple est de se ramener au corps $\mathbb{C}$ comme dans la démonstration de 4.3 et d'utiliser alors la théorie des groupes de Lie : les éléments de $S^{\wedge}(\underline{g}^{\star})$ considérés sont dans ce cas des fonctions analytiques dans un voisinage de 0 de $\underline{g}$.

Considérons la "fonction" $x \mapsto \det \frac{\exp adx - 1}{adx}$. C'est une série formelle de terme constant 1 . Soit $j_d \in S^{\wedge}(\underline{g}^{\star})$ la racine carrée de cette série dont le terme constant est 1 . Définissons de même j_g à partir de la "fonction" $x \mapsto \det \frac{1-\exp(-adx)}{adx}$.
Ces fonctions sont centrales, donc

(6) $$[W(x),j_g]=[W(x),j_d]=0 .$$

Notons trad l'élément $x \mapsto tr(ad(x))$ de $\underline{g}^{\star}$. On a $j_d^2 j_g^{-2}=\det \exp adx$, donc

(7) $$j_d j_g^{-1}=\exp(\tfrac{1}{2}\, trad) .$$

La forme linéaire trad s'annule sur $[\underline{g},\underline{g}]$, donc commute avec $[\underline{g},\underline{g}]$. Compte tenu des formules de 2.2 et 2.3, on a

(8) $$[L(x),trad]=[R(x),trad]=[x,trad]=trad\ x .$$

Si le corps de base est $\mathbb{C}$, et si $(q_1,..,q_n)$ est une base de $\underline{g}^{\star}$, la forme différentielle $j_d^2 dq_1 \dots dq_n$ est invariante par translation à droite, donc

$$0=L(x)(j_d^2 dq_1\dots dq_n)=2j_d(L(x)j_d)dq_1\dots dq_n+j_d^2(\operatorname{div} L(x))dq_1\dots dq_n .$$

Donc, compte tenu de (6) ,

(9) $$\operatorname{div} L(x)=-2j_d^{-1}[L(x),j_d]= -2j_d^{-1}[R(x),j_d] .$$

(On notera que, si $a\in S^{\wedge}(\underline{g}^{\star})$, et si $b\in S^{\wedge}(\underline{g}^{\star})\underline{g}$ est un champ de vecteurs, l'action sur a de l'opérateur différentiel b donne $[b,a]$) . De même

(10) $$\operatorname{div} R(x)= -2j_g^{-1}[R(x),j_g]= -2j_g^{-1}[L(x),j_g] .$$

On définit les antiautomorphismes $\perp$ et $\top$ de $A^{\wedge}(\underline{g})$ comme en 1.5 . L'action sur $S^{\wedge}(\underline{g}^{\star})$ de $\top$ est l'identité , et l'action de $\perp$ transforme une "fonction" $f(x)$ en la fonction $f(-x)$. En particulier

(11) $$j_g^{\perp}=j_d \qquad j_d^{\perp}=j_g .$$

En utilisant les formules de 2.2 et 2.3, on trouve facilement que $L(x)^{\perp}= -R(x)^{\top}$.
Compte tenu de 1.5,

(12) $$L(x)^{\perp}= -R(x)^{\top}=R(x)+\operatorname{div} R(x)$$

et de même

(13) $$R(x)^{\perp}=-L(x)^{\top}=L(x)+\operatorname{div} L(x) .$$

On en déduit

(14) $$W(x)^{\perp}=W(x)^{\top}=-W(x)-\operatorname{div} W(x)=-W(x)+trad\ x .$$

6.3. Pour tout $u \in U(\underline{g})$, on pose

(15) $L'(u)=j_d L(u) j_d^{-1}$ $R'(u)=j_g R(u) j_g^{-1}$.

En particulier, pour $x \in \underline{g}$,

(16) $L'(x)=L(x)-j_d^{-1}[L(x),j_d]=L(x)+\frac{1}{2}\operatorname{div} L(x)$

(17) $R'(x)=R(x)-j_g^{-1}[R(x),j_g]=R(x)+\frac{1}{2}\operatorname{div} R(x)$,

d'après (9) et (10) . Posons

(18) $W'(x)=L'(x)-R'(x)=W(x)+\frac{1}{2}\operatorname{div} W(x)=W(x)-\frac{1}{2}\operatorname{trad} x$.

Nous allons voir que les résultats du §2 s'étendent à condition d'y remplacer L,R,W par L',R',W' .

Il est immédiat que l'application L'(resp.R') est un homomorphisme (resp.antihomomorphisme) de $U(\underline{g})$ dans $A^{\wedge}(\underline{g})$. L'application W' est un homomorphisme d'algèbres de Lie. D'après (8), on a, pour tout $x \in \underline{g}$,

$$[L(x),\exp\frac{1}{2}\operatorname{trad}]=(\exp\frac{1}{2}\operatorname{trad})\frac{1}{2}(\operatorname{trad} x) ,$$

donc, pour tout $y \in \underline{g}$,

$$[L(x),\exp(-\frac{1}{2}\operatorname{trad})R(y)\exp(\frac{1}{2}\operatorname{trad})]$$

$$= -(\exp(-\frac{1}{2}\operatorname{trad}))\frac{1}{2}(\operatorname{trad} x)R(y)\exp(\frac{1}{2}\operatorname{trad})+(\exp(-\frac{1}{2}\operatorname{trad}))R(y)(\exp\frac{1}{2}\operatorname{trad})\frac{1}{2}(\operatorname{trad} x)$$

$=0$.

Ainsi $L(x)$ commute à $j_g j_d^{-1} R(y) j_d j_g^{-1}$, donc $L'(x)$ commute à $R'(y)$. On en déduit que

(19) $$[L'(u),R'(v)]=0 \qquad \text{pour } u,v \in U(\underline{g}) .$$

Le lemme 2.5 s'étend trivialement. Si $x \in \underline{g}$, on a

$$L'(x)^{\perp}=(j_d^{-1})^{\perp}L(x)^{\perp}(j_d)^{\perp}=j_g^{-1}(R(x)+\operatorname{div} R(x))j_g$$

$$=R(x)+j_g^{-1}[R(x),j_g]+\operatorname{div} R(x)=R(x)+\frac{1}{2}\operatorname{div} R(x)=R'(x) ;$$

(on a utilisé (15),(11),(12),(10),(17)) . Donc

(20) $$L'(u)^{\perp}=R'(u) \qquad \text{pour tout } u \in U(\underline{g}) .$$

D'autre part,

$$L'(x)^{\top}=L(x)^{\top}+\frac{1}{2}\operatorname{div} L(x)= -L(x)-\operatorname{div} L(x)+\frac{1}{2}\operatorname{div} L(x)$$

$$= -L(x)-\frac{1}{2}\operatorname{div} L(x)= -L'(x)=L'(-x)=L'(x^{\top}) ;$$

(on a utilisé (16),(13)) ; de même, $R'(x)^{\top}=R'(x^{\top})$. Donc

(21) $$L'(u^{\top})=L'(u)^{\top} \qquad R'(u^{\top})=R'(u)^{\top} \qquad \text{pour tout } u \in U(\underline{g}) .$$

6.4. On peut alors étendre les résultats du §3 . Soit $\Lambda'(\underline{g})$ (resp. $P'(\underline{g})$) l'idéal à gauche (resp. à droite) de $A^{\wedge}(\underline{g})$ engendré par les $W'(x)$ pour $x \in \underline{g}$. Alors $\Lambda'(\underline{g})$ et $P'(\underline{g})$ sont invariants par l'action adjointe de $\underline{g}$. Pour tout $u \in U(\underline{g})$, on a $R'(u)-L'(u) \in \Lambda'(\underline{g}) \cap P'(\underline{g})$ (même démonstration qu'en 3.2). D'après (14) et (18), on a

$$W'(x)^{\perp}=W'(x)^{\top}= -W(x)+\operatorname{trad} x-\frac{1}{2}\operatorname{trad} x$$

donc

(22) $$W'(x)^{\perp}=W'(x)^{\top}= -W'(x) .$$

Par suite, $\Lambda'(\underline{g})^{\perp}=\Lambda'(\underline{g})^{\tau}=P'(\underline{g})$. Alors, si J est un idéal de $S(\underline{g})$, on voit comme en 3.3 que

(23) $$L'^{-1}(A^{\wedge}(\underline{g})J+\Lambda'(\underline{g}))=R'^{-1}(A^{\wedge}(\underline{g})J+\Lambda'(\underline{g}))$$
$$=L'^{-1}(JA^{\wedge}(\underline{g})+P'(\underline{g}))=R'^{-1}(JA^{\wedge}(\underline{g})+P'(\underline{g})) .$$

Cet ensemble est un idéal bilatère de $U(\underline{g})$ qu'on notera $\mu'(J)$. On a

(24) $$\mu'(J^{\tau})=(\mu'(J))^{\tau} .$$

Soit Γ le groupe adjoint algébrique de $\underline{g}$. Si $J'=\sum_{\gamma\in\Gamma}\gamma J$, on voit comme en 3.3 que $A^{\wedge}(\underline{g})J'+\Lambda'(\underline{g})=A^{\wedge}(\underline{g})J+\Lambda'(\underline{g})$ et $\mu'(J')=\mu'(J)$.

6.5. Soit ω une Γ-orbite dans $\underline{g}^{*}$. Si $f\in\omega$ et si $x\in\underline{g}^{f}$, ad x annule une forme bilinéaire alternée non dégénérée sur $\underline{g}/\underline{g}^{f}$. Si de plus les éléments de ω sont <u>réguliers</u> , $\underline{g}^{f}$ est commutative, donc trad $x=0$. Ainsi, $\underline{g}^{f}$ est contenu dans l'idéal $\underline{k}=\mathrm{Ker}(\mathrm{trad})$ de $\underline{g}$, qui est de codimension 1 ou 0 . D'après [5] , prop.1.1 (du moins pour $k=\mathbb{R}$) , on a alors $\omega+k(\mathrm{trad})=\omega$. Il résulte de là que <u>l'idéal</u> $J(\omega)$ <u>défini par</u> ω <u>est engendré par son intersection avec</u> $S(\underline{k})$.

Pour tout $\lambda\in k$, soit e_{λ} l'élément $\exp(\lambda\ \mathrm{trad})$ de $S^{\wedge}(\underline{g}^{*})$. Alors l'automorphisme intérieur de $A^{\wedge}(\underline{g})$ défini par e_{λ} est l'identité sur $\underline{k}$ donc sur $S(\underline{k})$. En particulier, $e_{\lambda}J(\omega)e_{\lambda}^{-1}=J(\omega)$. D'autre part, e_{λ} commute avec les $W(x)$ donc avec les $W'(x)$. Par suite, $\mu'(J(\omega))=L'^{-1}(A^{\wedge}(\underline{g})e_{\lambda}J(\omega)e_{\lambda}^{-1}+e_{\lambda}\Lambda'(\underline{g})e_{\lambda}^{-1})$.

Soit $\lambda'\in k$ et posons, pour tout $u\in U(\underline{g})$,

(25) $$L''(u)=e_{\lambda}^{-1}L'(u)e_{\lambda} \qquad R''(u)=e_{\lambda'}^{-1}R'(u)e_{\lambda'} .$$

Alors, d'après ce qui précède,

(26) $$\mu'(J(\omega))=L''^{-1}(A^{\wedge}(\underline{g})J(\omega)+\Lambda'(\underline{g}))=R''^{-1}(A^{\wedge}(\underline{g})J(\omega)+\Lambda'(\underline{g})) .$$
$$=L''^{-1}(J(\omega)A^{\wedge}(\underline{g})+P'(\underline{g}))=R''^{-1}(J(\omega)A^{\wedge}(\underline{g})+P'(\underline{g})) .$$

En particulier, prenons $\lambda=\frac{1}{4}$, $\lambda'=-\frac{1}{4}$, d'où $e_{\lambda}=(j_d j_g^{-1})^{\frac{1}{2}}=e_{\lambda'}^{-1}$.

Posons

(27) $$j=(j_g j_d)^{1/2} .$$

On a alors, d'après (15) et (25),

(28) $$L''(u)=jL(u)j^{-1} \qquad R''(u)=jR(u)j^{-1} .$$

6.6. Dans le cas des orbites non régulières, il faut probablement considérer un homomorphisme $U(\underline{g})\longrightarrow A^{\wedge}(\underline{g})$ de la forme $u\longmapsto gL(u)g^{-1}$, où g est un élément inversible de $S^{\wedge}(\underline{g}^{*})$ qui ne dépend que de la nappe de l'orbite (lorsque $\underline{g}$ n'est pas résoluble, il faut même probablement prendre g dans une extension algébrique du corps des fractions de $S^{\wedge}(\underline{g}^{*})$) . Montrons seulement, sur un exemple, que la méthode permet d'associer des idéaux primitifs raisonnables de $U(\underline{g})$ à certaines orbites <u>non polarisables</u>. Les calculs ne seront qu'esquissés.

6.7. Soit $\underline{s}=\underline{sl}(2,\mathbb{C})$ avec sa base habituelle (e,h,f) . Soit $\underline{h}$ l'algèbre de Heisenberg de base (u,v,w) telle que $[u,v]=w$, $[u,w]=[v,w]=0$. Faisons opérer $\underline{s}$

dans $\underline{h}$ par

$$[e,u]=0 \ , \ [e,v]=u \ , \ [h,u]=u \ , \ [h,v]=-v \ , \ [f,u]=v \ , \ [f,v]=0 \ , \ [\underline{s},w]=0 \ .$$

Soit $\underline{g}$ le produit semi-direct correspondant. Il a pour base (e,h, f,u,v,w) . C'est une algèbre de Lie algébrique unimodulaire. Soit Γ le groupe adjoint de $\underline{g}$.

Soit A (resp.B) l'ensemble des points de $\underline{g}^\star$ où s'annulent les éléments u^2-2ew , $uv+hw$, v^2+2fw (resp.u,v,w) de $S(\underline{g})$. On a $A\supset B$. La variété A a deux composantes irréductibles de dimension 3 ; l'une est B , l'autre est l'adhérence de A-B . Les éléments de A-B sont les éléments non polarisables de $\underline{g}^\star$; l'ensemble A-B est réunion d'une famille $(\omega_\lambda)_{\lambda\in\mathbb{C}-\{0\}}$ de Γ-orbites fermées de dimension 2 . L'idéal J_λ de $S(\underline{g})$ correspondant à ω_λ est engendré par $u^2-2\lambda e$, $uv+\lambda h$, $v^2+2\lambda f$, $w-\lambda$. Dans $\underline{g}^\star/\Gamma$, soit P l'image de l'ensemble des éléments polarisables de $\underline{g}^\star$.

Dans l'algèbre localisée $U(\underline{g})_w$, on considère les éléments suivants :

$$e-\frac{u^2}{2w} \ , \ h+\frac{uv+vu}{2w} \ , \ f+\frac{v^2}{2w} \ , \ u, \ \frac{v}{w} \ , \ w \ .$$

En calculant leurs crochets, on constate qu'il existe un isomorphisme θ de $U(\underline{s})\otimes A_1[p,q]\otimes k[w]_w$ (où $A_1[p,q]$ désigne l'algèbre de Weyl de générateurs canoniques p,q) sur $U(\underline{g})_w$ tel que

$$\theta(e\otimes 1\otimes 1)=e-\frac{u^2}{2w} \ , \quad \theta(h\otimes 1\otimes 1)=h+\frac{uv+vu}{2w} \ , \quad \theta(f\otimes 1\otimes 1)=f+\frac{v^2}{2w} \ ,$$

$$\theta(1\otimes p\otimes 1)=u \qquad \theta(1\otimes q\otimes 1)=\frac{v}{w} \qquad \theta(1\otimes 1\otimes w)=w \ .$$

Cela permet de calculer assez facilement l'ensemble Prim $U(\underline{g})$. Soit Primc $U(\underline{g})$ l'ensemble des idéaux primitifs complètement premiers de $U(\underline{g})$.

On peut appliquer la méthode des orbites à partir de P . On obtient ainsi une bijection de P sur une partie P' de Primc $U(\underline{g})$. Le complémentaire Primc $U(\underline{g})$-P' est un ensemble d'idéaux I_λ ($\lambda\in\mathbb{C}-\{0\}$) définis de la manière suivante : Si $\lambda\in\mathbb{C}-\{0\}$, I_λ est l'idéal bilatère de $U(\underline{g})$ engendré par les éléments

$$u^2-2\lambda e \ , \ uv+vu+2\lambda h \ , \ v^2+2\lambda f \ , \ w-\lambda \ .$$

Le quotient $U(\underline{g})/I_\lambda$ est engendré par les images canoniques de u,v , et est une algèbre de Weyl; en particulier, les I_λ sont des idéaux bilatères maximaux de $U(\underline{g})$.

On pense irrésistiblement que les I_λ devraient être associés aux orbites ω_λ . Comme cela semble impossible par la méthode des polarisations, nous allons utiliser la méthode du présent article.

6.8. Nous noterons (x,y,z,a,b,c) la base de $\underline{g}^\star$ duale de (e,h,f,u,v,w) . On a $[\lambda_1 e+\lambda_2 h+\lambda_3 f+\lambda_4 u+\lambda_5 v+\lambda_6 w,e]=2\lambda_2 e-\lambda_3 h-\lambda_5 u$, d'où $W(e)=2ye-zh-bu$

et de même

$$W(h)=-2xe+2zf-au+bv, \quad W(f)=xh-2yf-av \ ,$$

$$W(u)=yu+zv-bw \ , \quad W(v)=xu-yv+aw \ , \quad W(w)=0 \ .$$

Il sera commode de poser $\delta=\sqrt{y^2+xz}$ (cette fonction n'interviendra, sauf tout à fait à la fin, que par son carré). Soit T la matrice de $ad_{\underline{g}}(xe+yh+zf+au+bv+cw)$ (pour ne pas multiplier les notations, on considère provisoirement x,y,...,c comme les coordonnées d'un point fixé de $\underline{g}$) . Soit $S=T/(1-e^{-T})$. En général, T est semi-simple, de valeurs propres $\pm 2\delta, \pm\delta, 0, 0$. Soient $\varphi(\delta), \psi(\delta)$ les fonctions (ou séries formelles en x,y,z) définies par

$$1+\varphi\delta^2+\psi\delta^4=\frac{\delta}{2}\coth\frac{\delta}{2} \qquad 1+4\varphi\delta^2+16\psi\delta^4=\delta\coth\delta .$$

Alors le polynôme $1+\varphi X^2+\psi X^4$ prend les mêmes valeurs que

$$\frac{X}{1-e^{-X}}-\frac{X}{2}=\frac{X}{2}\coth\frac{X}{2}$$

en $0, \pm\delta, \pm 2\delta$, donc

$$S=1+\frac{1}{2}T+\varphi T^2+\psi T^4$$

ce qui permet le calcul de S . On trouve alors :

$$L(e)=e+\frac{1}{2}W(e)+(\varphi+4\delta^2\psi)(2yW(e)-zW(h)-bW(u))+3\psi(za-yb)(yW(u)+zW(v))$$

$$L(h)=h+\frac{1}{2}W(h)+(\varphi+4\delta^2\psi)(-2xW(e)+2zW(f)+bW(v)-aW(u))$$
$$+3\psi(y^2a-xza+2xyb)\,W(u)+3\psi(2yza+xzb-y^2b)W(v)$$

$$L(f)=f+\frac{1}{2}W(f)+(\varphi+4\delta^2\psi)(xW(h)-2yW(f)-aW(v))+3\psi(xb+ya)(xW(u)-yW(v))$$

$$L(u)=u+\frac{1}{2}W(u)+(\varphi+\delta^2\psi)(yW(u)+zW(v))$$

$$L(v)=v+\frac{1}{2}W(v)+(\varphi+\delta^2\psi)(xW(u)-yW(v))$$

$$L(w)=w .$$

Calculant modulo $\Lambda'(\underline{g})$, on en déduit

$$L(2\lambda e-u^2)\equiv 2\lambda e-u^2+\lambda z(\varphi+\delta^2\psi)$$
$$L(2\lambda h+uv+vu)\equiv 2\lambda h+2uv+2\lambda y(\varphi+\delta^2\psi)$$
$$L(2\lambda f+v^2)\equiv 2\lambda f+v^2+\lambda x(\varphi+\delta^2\psi)$$
$$L(w-\lambda)=w-\lambda .$$

Posons $g=\mathrm{sh}\,\frac{\delta}{2}/\frac{\delta}{2}\in S\hat{}(\underline{g}^\star)$. La dérivée logarithmique de g par raport à δ est $\frac{1}{2}\coth\frac{\delta}{2}-\frac{1}{\delta}=\delta(\varphi+\delta^2\psi)$. L'élément g est Γ-invariant et commute à u,v,w . On a

(29) $$[2\lambda e-u^2,g^{-1}]=2\lambda[e,g^{-1}]=-2\lambda g^{-2}\frac{\partial g}{\partial x}=-2\lambda g^{-1}\delta(\varphi+\delta^2\psi)\frac{\partial\delta}{\partial x}$$
$$=-2\lambda g^{-1}\delta(\varphi+\delta^2\psi)\frac{z}{2\delta}=-\lambda zg^{-1}(\varphi+\delta^2\psi) .$$

Comme g est Γ-invariant, g commute à $W_{\underline{g}}(\underline{g})$ (cf.2.5) , donc

$$gL(2\lambda e-u^2)g^{-1}\equiv g(2\lambda e-u^2+\lambda z(\varphi+\delta^2\psi))g^{-1}$$
$$=2\lambda e-u^2+g[2\lambda e-u^2,g^{-1}]+\lambda z(\varphi+\delta^2\psi)$$

soit, d'après (29),

$$gL(2\lambda e-u^2)g^{-1}\equiv 2\lambda e-u^2\in J_\lambda .$$

On trouve de même

$$gL(2\lambda h+uv+vu)g^{-1}\equiv 2\lambda h+2uv\in J_\lambda ,$$

$gL(2\lambda f+v^2)g^{-1} \equiv 2\lambda f+v^2 \in J_\lambda$

$gL(w-\lambda)g^{-1} = w-\lambda \in J_\lambda$.

Soit I'_λ l'ensemble des $t\in U(\underline{g})$ tels que $gL(t)g^{-1}\in A^\frown(\underline{g})J_\lambda + \Lambda'(\underline{g})$. On a $2\lambda e-u^2, 2\lambda h+uv+vu, 2\lambda f+v^2, w-\lambda \in I'_\lambda$. L'idéal bilatère I'_λ contient donc I_λ . Montrons que $I'_\lambda = I_\lambda$. Comme I_λ est maximal, il suffit de prouver que $I'_\lambda \neq U(\underline{g})$, donc que $A^\frown(\underline{g})J_\lambda + \Lambda'(\underline{g}) \neq A^\frown(\underline{g})$.

Soit K le corps des fractions de $S^\frown(\underline{g}^\star)$. L'algèbre $A^\frown(\underline{g})$ opère naturellement dans K . Considérons l'élément suivant de K :

$$\psi_\lambda = \lambda\left(c+\frac{b^2x+2aby-a^2z}{y^2+xz}\right) .$$

Adjoignons à K un élément transcendant que nous notons $\exp\psi_\lambda$. L'action de $A^\frown(\underline{g})$ sur K se prolonge à $K(\exp\psi_\lambda)$; on pose :

$$e.(\exp\psi_\lambda) = \frac{\partial}{\partial x}(\exp\psi_\lambda) = (\exp\psi_\lambda)\frac{\partial\psi_\lambda}{\partial x} \text{ , etc .}$$

(L'existence de $K(\exp\psi_\lambda)$ et de l'action de $A^\frown(\underline{g})$ sur $K(\exp\psi_\lambda)$ avec les propriétés indiquées résulte de [8], p.407, exerc.2a ; comme me l'a indiqué E.R.Kolchin, la démonstration de ce résultat est analogue à celle qui est esquissée pour l'exerc. 1, loc. cit.). Enfin, adjoignons à $K(\exp\psi_\lambda)$ l'élément algébrique $\delta=(y^2+xz)^{1/2}$. L'action de $A^\frown(\underline{g})$ se prolonge naturellement à $K(\exp\psi_\lambda,\delta)$. Posons $\varphi_\lambda = \delta^{-1}\exp\psi_\lambda$. Le calcul montre que φ_λ est annulé par

$$\frac{\partial^2}{\partial a^2} - 2\lambda\frac{\partial}{\partial x} = u^2-2\lambda e \text{ , } \frac{\partial^2}{\partial a\partial b} + \lambda\frac{\partial}{\partial y} = uv+\lambda h \text{ , } \frac{\partial^2}{\partial b^2} + 2\lambda\frac{\partial}{\partial z} = v^2+2\lambda f \text{ ,}$$

$$\frac{\partial}{\partial c} - \lambda = w-\lambda, 2y\frac{\partial}{\partial x} - z\frac{\partial}{\partial y} - b\frac{\partial}{\partial a} = W(e),\dots,W(v),W(w) .$$

Cela prouve bien que $A^\frown(\underline{g})J_\lambda + \Lambda'(\underline{g}) \neq A^\frown(\underline{g})$.

7. Quelques remarques sur les opérateurs L(u) .

Soit $\underline{g}$ une algèbre de Lie de dimension finie sur k .

7.1. Lemme. Soient $x\in\underline{g}$, $(e_1,\dots,e_n)$ une base de $\underline{g}$. Il existe des éléments $\varphi_1,\dots,\varphi_n$ de $S^\frown(\underline{g}^\star)$ (et même de $S(\underline{g}^\star)$ si $\underline{g}$ est nilpotente) tels que

$$L(x) = x+\varphi_1 W(e_1)+\dots+\varphi_n W(e_n) .$$

En effet, la "valeur" de $L(x)-x$ en $y\in\underline{g}$ est

$$\textstyle\sum_{r\geq 1} b_r(\mathrm{ad}\,y)^r x = [y, \sum_{r\geq 1} b_r(\mathrm{ad}\,y)^{r-1}x]$$
$$= [y, \varphi_1(y)e_1+\dots+\varphi_n(y)e_n]$$

où $\varphi_1,\dots,\varphi_n$ sont des éléments de $S^\frown(\underline{g}^\star)$, et même de $S(\underline{g}^\star)$ si $\underline{g}$ est nilpotente. Autrement dit,

$$L(x)(y) = x+\varphi_1(y)W(e_1)(y)+\dots+\varphi_n(y)W(e_n)(y) .$$

7.2. Lemme. Soient $x \in \underline{g}$, $v \in U(\underline{g})$, $u = xv \in U(\underline{g})$. Avec les notations de 7.1, on a

$$L(u) = xL(v) + \sum_{i=1}^{n} \varphi_i L([e_i, v]) + \sum_{i=1}^{n} \varphi_i L(v) W(e_i).$$

En effet,

$$L(u) = L(x)L(v) = xL(v) + \sum_{i=1}^{n} \varphi_i W(e_i) L(v)$$

et

$$W(e_i)L(v) = L(v)W(e_i) + [W(e_i), L(v)]$$
$$= L(v)W(e_i) + [L(e_i), L(v)] = L(v)W(e_i) + L([e_i, v]).$$

7.3. Lemme. Soient $(e_1, \ldots, e_n)$ une base de $\underline{g}$, et $u = \sum_{|\alpha| \leq p} \lambda_\alpha e_1^{\alpha_1} \ldots e_n^{\alpha_n} \in U(\underline{g})$ où les λ_α appartiennent à k. Alors, dans $A\hat{}(\underline{g})$, on a

$$L(u) = \sum_{|\alpha| = p} \lambda_\alpha e_1^{\alpha_1} \ldots e_n^{\alpha_n}$$
$$+ \sum_{|\beta| < p} \psi_\beta e_1^{\beta_1} \ldots e_n^{\beta_n} + \sum_{|\gamma| < p, 1 \leq i \leq n} \omega_{\gamma,i} e_1^{\gamma_1} \ldots e_n^{\gamma_n} W(e_i)$$

où les ψ_β et les $\omega_{\gamma,i}$ appartiennent à $S\hat{}(\underline{g}^*)$ (et même à $S(\underline{g}^*)$ si $\underline{g}$ est nilpotente).

C'est clair quand $p=0$. Supposons le lemme établi pour $p-1$. Il suffit d'envisager le cas où $u = xv$, avec x dans $\underline{g}$ et v de filtration $\leq p-1$. Alors les $[e_i, v]$ sont de filtration $\leq p-1$. Utilisons le lemme 7.2 et l'hypothèse de récurrence. Le terme $\sum_{i=1}^{n} \varphi_i L([e_i, v])$ est de la forme $\sum_{|\beta| < p} \psi_\beta e_1^{\beta_1} \ldots e_n^{\beta_n}$ avec des ψ_β dans $S\hat{}(\underline{g}^*)$. Le terme $\sum_{i=1}^{n} \varphi_i L(v) W(e_i)$ est de la forme

$$\sum_{|\gamma| < p, 1 \leq i \leq n} \omega_{\gamma,i} e_1^{\gamma_1} \ldots e_n^{\gamma_n} W(e_i)$$

avec des $\omega_{\gamma,i}$ dans $S\hat{}(\underline{g})$. Si $v = \sum_{|\delta| \leq p-1} \mu_\delta e_1^{\delta_1} \ldots e_n^{\delta_n}$ dans $U(\underline{g})$, on a dans $A\hat{}(\underline{g})$

$$xL(v) = \sum_{|\delta| = p-1} \mu_\delta x e_1^{\delta_1} \ldots e_n^{\delta_n}$$
$$+ \sum_{|\beta| < p-1} \psi_\beta x e_1^{\beta_1} \ldots e_n^{\beta_n} + \sum_{|\gamma| < p-1, 1 \leq i \leq n} \omega_{\gamma,i} x e_1^{\gamma_1} \ldots e_n^{\gamma_n} W(e_i)$$

avec des $\psi_\beta, \omega_{\gamma,i}$ dans $S\hat{}(\underline{g}^*)$.

Or on a dans $S(\underline{g})$

$$\sum_{|\alpha| = p} \lambda_\alpha e_1^{\alpha_1} \ldots e_n^{\alpha_n}$$
$$= \sum_{|\delta| = p-1} \mu_\delta x e_1^{\delta_1} \ldots e_n^{\delta_n}$$

d'où le lemme.

7.4. Proposition. Soit $u \in U(\underline{g})$ un élément de filtration p. Soit u' l'image canonique de u dans l'ensemble des éléments de $S(\underline{g})$ homogènes de degré p. Alors, modulo $\Lambda(\underline{g})$, $L(u)$ est congru à un opérateur différentiel d'ordre p dont la partie principale est u'.

Cela résulte aussitôt de 7.3.

7.5. Corollaire. Soient $\underline{g}$ une algèbre de Lie semi-simple réelle, $u \in U(\underline{g})$ l'élément de Casimir, $s \in S(\underline{g})$ l'image de u par la symétrisation, j la fonction $x \mapsto (\det \operatorname{sh} \frac{\operatorname{ad} x}{2} / \frac{\operatorname{ad} x}{2})^{1/2}$ au voisinage de 0 dans $\underline{g}$. On a $jL(u)j^{-1} - s \in \Lambda(\underline{g})$.
(Ce résultat a été aussi obtenu par Kashiwara et Vergne).

D'après 7.4, on a $L(u) = s + t + w$, où t est un opérateur différentiel d'ordre ≤ 1 et $w \in \Lambda(\underline{g})$ (toutes les séries formelles qui interviennent ici ont un rayon de convergence >0). Donc $jL(u)j^{-1} = s + t' + w'$, où t' est d'ordre ≤ 1 et $w' \in \Lambda(\underline{g})$ (car j commute à $W(\underline{g})$). D'après [10], p.113, prop.8.2.4.2, $jL(u)j^{-1} - s$ annule toute fonction C^∞ localement invariante sur $\underline{g}$ au voisinage de 0. Appliquant cela d'abord à la fonction 1, on voit que $t'(1) = 0$ donc que t' est un champ de vecteurs. Comme ce champ de vecteurs annule toute fonction C^∞ localement invariante au voisinage de 0, on a $t' \in \Lambda(\underline{g})$ d'après [4], th.2.6.

7.6. En utilisant le théorème de Hamilton-Cayley, on voit que les fonctions $\varphi_1, \ldots, \varphi_n$ de 7.1 peuvent être prises dans la sous-algèbre de $S^\wedge(\underline{g}^*)$ engendrée par $S(\underline{g}^*)$ et les éléments invariants de $S^\wedge(\underline{g}^*)$. Il en est donc de même des fonctions $\psi_\beta, \omega_{\gamma,i}$ qui interviennent dans 7.3.

◊ ◊ ◊ ◊ ◊ ◊ ◊ ◊

Bibliographie

[1] N.Bourbaki, Variétés différentielles et analytiques, §§8-15, Paris, Hermann, 1971.

[2] J.Dixmier, Algèbres enveloppantes, Paris, Gauthier-Villars, 1974.

[3] J.Dixmier, Sur les distributions invariantes par un groupe nilpotent, C.R. Acad.Sci.Paris, t.285, 1977, p.7 -10.

[4] J.Dixmier, Champs de vecteurs adjoints sur les groupes et algèbres de Lie semi-simples, à paraître.

[5] M.Duflo, Caractères des groupes et des algèbres de Lie résolubles, Ann.scient.Ec.Norm.Sup, t.3, 1970, p.23-74.

[6] M.Duflo, Sur les extensions des représentations irréductibles des groupes de Lie nilpotents, Ann.scient.Ec.Norm.Sup., t.5, 1972, p.71-120.

[7] M.Duflo, Opérateurs différentiels bi-invariants sur un groupe de Lie, Ann. scient.Ec.Norm.Sup., t.10, 1977, p.265-288.

[8] E.R.Kolchin, Differential algebra and algebraic groups, Academic Press, New-York and London, 1973.

[9] L.Schwartz, Théorie des distributions, nouvelle édition, Paris, Hermann, 1966.

[10] G.Warner, Harmonic analysis on semi-simple Lie groups, II, Berlin-Heidelberg-New-York, Springer, 1972.

POLYNOMES DE VOGAN POUR SL(n,ℂ).

Michel Duflo (*)

Summary. In [13] Vogan associates to an irreducible Harish-Chandra module for a real semi-simple Lie group a polynomial on a Cartan subalgebra. I prove that in the case of the group SL(n,ℂ), some conjectures made by Vogan on these polynomials are true. The proof uses some of the Joseph's ideas in [6].

Résumé. Dans [13] Vogan associe à un module de Harish-Chandra irréductible pour un groupe de Lie semi-simple réel un polynôme sur une sous-algèbre de Cartan. Je démontre que, dans le cas du groupe SL(n,ℂ), certaines conjectures faites par Vogan sur ces polynômes sont vraies. La démonstration utilise certaines des idées de Joseph [6].

(*) Université Paris 7 et Ecole Polytechnique.

I. POLYNOMES DE VOGAN. Je rappelle ici les résultats de [13] indispensables à la compréhension de ce qui suit.

Soit $\underline{g}$ une algèbre de Lie de dimension finie sur un corps k, et soit V un module de type fini sur l'algèbre enveloppante $U(\underline{g})$. On note $U_n(\underline{g})$ $(n \in \mathbb{N})$ la filtration standard de $U(\underline{g})$. On choisit un sous-espace de dimension finie V_o de V engendrant V. Pour les grandes valeurs de n, dim $U_n(\underline{g})V_o$ est un polynôme en n. Son terme dominant ne dépend pas du choix de V_o. Il est de la forme $e(V)\, n^{d(V)}/d(V)\,!$, où $d(V) \in \mathbb{N}$ est la "dimension de Gelfand-Kirillov" de V et où $e(V) \in \mathbb{N}^*$ est la "multiplicité" de V. Suivant [13], on définit $e_d(V)$ pour tout entier $d \geq d(V)$ en posant $e_{d(V)}(V) = e(V)$ et $e_d(V) = 0$ si $d > d(V)$. (Pour tout ceci, voir par exemple les références rappelées dans [13]).

Soit G un groupe de Lie semi-simple réel. On note $\underline{g}_o$ son algèbre de Lie, $\underline{g}$ le compléxifié de $\underline{g}_o$. Pour simplifier, on suppose que G est le sous-groupe analytique d'algèbre $\underline{g}_o$ du groupe simplement connexe $G_{\mathbb{C}}$ d'algèbre $\underline{g}$. On choisit un sous-groupe compact maximal K de G et une sous-algèbre de Cartan $\underline{h}$ de $\underline{g}$. On note R le système de racines de $\underline{g}$ par rapport à $\underline{h}$, on choisit un système R^+ de racines positives, on note W le groupe de Weyl, $s_\alpha \in W$ la symétrie correspondant à une racine α, P le réseau des poids, $\check{\alpha} \in \underline{h}$ la coracine correspondant à une racine α. Soit $\underline{h}^*$ le dual de $\underline{h}$. Si $p \in \underline{h}^*$, on note R_p le système de racines formé des $\alpha \in R$ tels que $\langle \check{\alpha}, p \rangle \in \mathbb{Z}$, $R_p^+ = R^+ \cap R_p$, $B_p \subset R_p^+$ le système de racines simples correspondant, W_p de sous-groupe de W formé des $w \in W$ tels que $wp - p \in \mathbb{Z}R$, χ_p le caractère du centre $Z(\underline{g})$ de $U(\underline{g})$ défini par Harish-Chandra.

Rappelons que l'on dit que p est dominant si $\langle p, \check{\alpha} \rangle \geq 0$ pour tout $\alpha \in R_p^+$, et régulier si $\langle p, \check{\alpha} \rangle \neq 0$ pour tout $\alpha \in R$.

Dans toute la suite de cet article, on fixe un élément $p \in \underline{h}^*$ dominant et régulier. Une "famille cohérente de g-K-modules (*) définie sur p + P" est une fonction Ⓜ définie sur p + P telle que :

(i) pour tout $\mu \in p + P$, Ⓜ(μ) est une combinaison linéaire finie à coefficients dans $\mathbb{Z}$ de classes d'équivalences de $\underline{g}$-K-modules simples de caractère infinitésimal χ_μ ;

(*) Dans ce texte nous dirons "$\underline{g}$-K-module" pour "$\underline{g}$-K-module admissible".

(ii) pour tout $\underline{g}$-module de dimension finie F et pour tout $\mu \in p + P$ on a

(1) $$\Theta(\mu) \otimes F = \sum_{\gamma \in P} n(\gamma)\Theta(\mu+\gamma)$$

où $n(\gamma)$ est la multiplicité du poids γ dans F.

(On remarquera que l'opération de produit tensoriel par une représentation de dimension finie est bien définie dans l'ensemble des combinaisons à coefficients dans $\mathbb{Z}$ de $\underline{g}$-K-modules simples).

Etant donné X une combinaison linéaire à coefficients dans $\mathbb{Z}$ de $\underline{g}$-K-modules simples de caractère infinitésimal χ_p, il existe une, et une seule, famille cohérente Θ telle que $\Theta(p) = X$. Pour tout ceci, voir [14], [10] et [11]. Rappelons les résultats suivants (cf. [14] et [12] th.3.20) :

On suppose que $\Theta(p) = X$ est simple.

(i) Si $\mu \in p + P$ est dominant régulier, $\Theta(\mu)$ est simple.

(ii) Si $\mu \in p + P$ est dominant, $\Theta(\mu)$ est nul ou simple.

(iii) Soit $\alpha \in B_p$. Ou bien $\Theta(s_\alpha p) = -\Theta(p)$, et dans ce cas $\Theta(\mu) = 0$ pour tout $\mu \in p + P$ dominant tel que $<\mu, \check{\alpha}> = 0$. Ou bien $\Theta(s_\alpha p)$ est combinaison (non vide) à coefficients dans $\mathbb{N}^*$ de modules simples, et dans ce cas $\Theta(\mu) \neq 0$ pour tout $\mu \in p + P$, dominant, tel que $<\mu, \check{\alpha}> = 0$, et $<\mu, \beta> \neq 0$ pour tout $\beta \in B_p$, $\beta \neq \check{\alpha}$.

On note $\tau(X)$ l'ensemble des $\alpha \in B_p$ tels que $\Theta(s_\alpha p) = -\Theta(p)$. Cet ensemble ne dépend que de l'annulateur ker(X) de X dans $U(\underline{g})$ et est égal à l'ensemble $\tau(\ker(X))$ défini dans [1] et [2] (cf. [13]).

Soit Θ une famille cohérente. Posons $\Theta(p) = \Sigma n_i X_i$ où les X_i sont simples, et $d = \sup_{n_i \neq 0} d(X_i)$. Soit $\mu \in p + P$, et posons $\Theta(\mu) = m_i Y_i$, où les Y_i sont simples. On a $d(Y_i) \leq d$ si $m_i \neq 0$, de sorte que l'on peut poser $\theta(\mu) = \Sigma m_i e_d(Y_i)$. Vogan [13] a démontré que θ est la restriction à $p + P$ d'un élément, que nous noterons encore θ, de $S(\underline{h})$. De plus, θ est W-harmonique. Vogan conjecture que les assertions suivantes sont vraies :

(i) θ est W_p-harmonique.

(ii) θ est homogène.

(iii) d + degré de $\theta = \# R^+$.

Je prouve ci-dessous (Théoreme 1, § III c) qu'il en est bien ainsi lorsque G est un groupe complexe et que R_p est un produit de systèmes simples de type A_m. (Cette dernière condition est toujours vérifiée si $G = SL(n, \mathbb{C})$).

Terminons ce paragraphe par quelques exemples de polynômes θ. Ces exemples ne servent pas dans la démonstration du théorème. Dans les exemples ci-dessus $\Theta(p) = X$ est supposé simple.

<u>Exemple</u> 1. On suppose que $p \in P$ et que $X = \Theta(p)$ est le module simple de dimension finie de caractère infinitésimal χ_p. Dans ce cas, pour tout $\mu \in P$ dominant régulier $\Theta(\mu)$ est le module simple de dimension finie de caractère infinitésimal χ_μ. On a donc :

$d(X) = 0$, $\theta(\mu) = \prod_{\alpha \in R^+} \langle \mu, \check{\alpha} \rangle / \langle \rho, \check{\alpha} \rangle$ (où ρ est la demi-somme des racines positives).

<u>Exemple</u> 2. On suppose $\tau(X) = \emptyset$. On a alors $d(X) = \# R^+$ et θ est constant (cf. [13], où il est montré de plus que ceci ne peut arriver que si G est quasi-déployé).

<u>Exemple</u> 3. On suppose $\tau(X) = B_p$. On a alors $d(X) = \# R^+ - \# R^+_p$ et $\theta = c \prod_{\alpha \in R^+_p} \check{\alpha}$ (où c est une constante non nulle).

On remarquera que l'exemple 1 est le cas particulier de l'exemple 3, avec $p \in P$.

<u>Exemple</u> 4. On suppose $\# B_p = 2$. Compte tenu des exemples ci-dessus, il reste à examiner le cas où $\tau(X) = \{\alpha\}$, avec $\alpha \in B_p$. On a alors $d(X) = \# R^+ - 1$ et $\theta = c\check{\alpha}$, où c est une constante non nulle.

Nous démontrerons les assertions des exemples 3 et 4 au § III, b.

II. <u>POLYNOMES DE VOGAN POUR LA CATEGORIE $\underline{O}$.</u>

Ils se définissent comme plus haut, avec des simplifications techniques. On s'en sert ici comme outil pour calculer certains polynômes de Vogan pour les $\underline{g}$-K-modules.

La catégorie $\underline{O}$ de $\underline{g}$-modules est définie par exemple en [3]. Des modules typiques de cette catégorie sont pour tout $\mu \in \underline{h}^*$ le module de Verma $M(\mu)$ de plus haut poids $\mu - \rho$, et son quotient simple $L(\mu)$.

Comme en I, on fixe $p \in \underline{h}^*$ régulier et dominant. Une " famille cohérente de modules de la catégorie $\underline{O}$ définie sur $-p + P$" est une fonction Ψ définie sur $-p + P$ telle que :

(i) pour tout $\mu \in -\rho + P, \Psi(\mu)$ est combinaison linéaire finie à coefficients dans $\mathbb{Z}$ de modules $L(w\mu)(w \in W)$;

(ii) La formule analogue à la formule (1) est vraie.

Etant donné une telle famille cohérente, on lui associe comme en I un élément W-harmonique $\psi \in S(\underline{h})$, à propos duquel on fait les mêmes conjectures qu'en I.

Soit $w \in W$. Nous noterons Ψ_w la famille cohérente telle que $\Psi_w(-\rho)=L($ et ψ_w le polynôme correspondant.

Soit S un sous-ensemble de B_p. On pose $\pi_S = \prod_{\mathbb{Z} S \cap R_p^+} \alpha$ et on note w_S l'élément de plus grande longueur du sous-groupe de W_p engendré par les s_α, $\alpha \in S$.

Lemme 1 (cf. [4]). On a $d(L(-w_S P)) = \# R^+ - \# \mathbb{Z} S \cap R_p^+$ et $\psi_{w_S} = c\,\pi_S$ où c est une constante non nulle.

Démonstration. Si $\mu \in -\rho + P$ est anti-dominant (i.e. $-\mu$ est dominant) et régulier, on a $\Psi_w(\mu) = L(w\mu)$, comme il résulte de [3] par exemple.
On a, d'après [3] 2.23 :

$$\mathrm{ch}\, L(w_S\,\mu) = \sum_{w \in W_S} \det(w\,w_S)\, \mathrm{ch}\, M(w\,\mu),$$

(Le caractère des modules de la catégorie $\underline{O}$ est défini par exemple dans [3]).

Notons ρ_S la demi-somme des éléments de $R_p^+ \cap \mathbb{Z} S$. Notons $\underline{g}_S$ l'algèb de Lie réductive admettant $\underline{h}$ comme sous-algèbre de Cartan et les $\check{\alpha}$ $(\alpha \in S)$ comm coracines . Si $\mu \in -\rho + P$ est antidominant régulier, $w_S\,\mu - \rho_S$ est le poids dominant d'un $\underline{g}_S$-module simple que nous noterons $F_S(w_S\,\mu - \rho_S)$. Si on note $m(\nu)$ l multiplicité du poids $\nu \in \underline{h}^*$ dans $F_S(w_S\,\mu - \rho_S)$, on a d'après les formules d'Hermann Weyl :

$$\sum m(\nu) = \det(w_S)\; P_S(\mu) \,/\, P_S(\rho_S)$$

$$\sum m(\gamma) e^{\gamma} = \frac{e^{-\rho_S} \sum \det(w)\, e^{w\,w_S\,\mu}}{\prod_{R_p^+ \cap \mathbb{Z} S} (1 - e^{-\alpha})} .$$

Posons d'autre part $K = \dfrac{e^{-\rho+\rho_S}}{\prod\limits_{\substack{\alpha\in R^+ \\ \alpha\notin R_p^+\cap \mathbb{Z}S}} (1-e^{-\alpha})}$.

Comme on a $\quad ch\, M(\nu) = \dfrac{e^{-\rho+\nu}}{\prod\limits_{\alpha\in R^+}(1-e^{-\alpha})}$ pour tout $\nu\in \underline{h}^*$, on obtient :

$$\text{(2)} \qquad ch\, L(w_S\mu) = K\ (\Sigma\ m(\gamma)\, e^{\gamma})\ .$$

Pour poursuivre la démonstration, nous avons besoin d'un lemme. On pose $r = \# B$.

<u>Lemme 2</u>. Soit $\gamma\in \underline{h}^*$. Soit M un $\underline{g}$-module, quotient de $M(\gamma)$. Si $m = (m_1, m_2, \dots, m_r)$ est un multi-indice, on pose $|m| = m_1 + m_2 + \dots + m_r$. Pour tout $n\in\mathbb{N}$ on note M^n la somme des sous-espaces de M correspondant aux poids $\gamma - \rho - \sum_{I=1}^{r} m_i\alpha_i$, avec $|m|\le n$. (Ici $\alpha_1,\dots,\alpha_r$ est l'ensemble des racines simples). Il existe des constantes $a, b > 0$ ne dépendant que de $\underline{g}$ telles que, pour tout $n\in\mathbb{N}$ assez grand on ait :

$$a\, e(M)\, n^{d(M)} \le \dim M^n \le b\, e(M)\, n^{d(M)}$$

<u>Démonstration</u>. Notons s le plus grand des nombres $|b|$, où pour tout élément $\beta\in R^+$ on écrit $\beta = \Sigma\, b_i\alpha_i$, On a

$$M^n \subset U_n(\underline{g})\, e \subset M^{ns}$$

où e est un vecteur non nul de M de poids $\gamma - \rho$. Le lemme en résulte facilement.

<u>Suite de la démonstration du lemme 1</u>.

Posons $\quad d = \# R^+ - \# R_p^+ \cap \mathbb{Z}S$ et $K = \sum_{\nu\in \underline{h}^*} k(\nu)\, e^{\nu}$. Posons

$$\varphi(n) = \sum_{|m|\le n} k(\rho_S - \rho - \Sigma\, m_i\alpha_i), \text{ si } n\in\mathbb{N}\,.$$

On voit facilement qu'il existe des constantes $a', b' > 0$ telles que l'on ait

$$\text{(3)} \qquad a'\, n^d \le \varphi(n) \le b'\, n^d$$

pour tout $n\in\mathbb{N}$.

On pose $M = L(w_S\mu)$ et on emploie les notations du lemme 1. Tout poids ν de $F_S(w_S\mu - \rho_S)$ s'écrit $w_S\mu - \rho_S - \Sigma b_i^{\gamma} \alpha_i$ avec des constantes $b_i \in$ On a d'après (2)

$$\dim M^n = \sum_{\gamma} m(\gamma)\, \varphi(n - |b^{\gamma}|)$$

pour tout $n \in \mathbb{N}$, et donc, d'après (3), on a, pour tout $n \in \mathbb{N}$ assez grand

$$\frac{1}{2}\, a' \,\Sigma\, m(\nu)\, n^d \leq \dim M^n \leq 2b' \,\Sigma\, m(\nu)\, n^d .$$

Rappelons que $\Sigma\, m(\nu)$ a été calculé plus haut. En comparant avec le lemme 2, on obtient le lemme 1.

III. GROUPES COMPLEXES.

III. - a) Définition de certaines familles cohérentes.

Nous allons considérer le groupe $G_{\mathbb{C}}$ comme réel. Le compléxifié de l'algèbre de Lie réelle sous-jacente à $\underline{g}$ est isomorphe à $\underline{g} \times \underline{g}$. Les polynômes de Vogan seront donc des éléments de $S(\underline{h}) \otimes S(\underline{h})$.

On note $u \to {}^t u$ un anti-automorphisme de $U(\underline{g})$ tel que ${}^t H = H$ pour tou $H \in \underline{h}$, et $u \to \acute{u}$ l'anti-automorphisme de $U(\underline{g})$ tel que $X = -X$ pour tout $X \in \underline{g}$. On choisit dans $G_{\mathbb{C}}$ le sous-groupe compact maximal U dont le compléxifié de l'algèbre de Lie est la sous-algèbre $\underline{u}$ de $\underline{g} \times \underline{g}$ formée des $(X, -{}^t X)$ $(X \in \underline{g})$.

Soient $\mu, \nu \in \underline{h}^*$ tels que $\mu - \nu \in P$. On note $\mathcal{L}(\mu,\nu)$ le $\underline{g} \times \underline{g} - U$ - module formé des vecteurs $\underline{u}$-finis du dual de $M(-\mu) \otimes M(-\nu)$ (c'est la série principale de $G_{\mathbb{C}}$) et $\mathcal{V}(\mu,\nu)$ le sous-quotient simple de $\mathcal{L}(\mu,\nu)$ qui contient le U-type mini

Soit $w \in W_p$. Nous noterons $Ⓗ_w$ la famille cohérente de $\underline{g} \times \underline{g} - U$ - mod définie sur $(p + P) \times (p + P)$ telle que $Ⓗ_w(p,p) = \mathcal{V}(wp,p)$, et ζ_w l'élément correspondant de $S(\underline{h}) \otimes S(\underline{h})$.

<u>Lemme 4</u>. Soient μ, $\nu \in p + P$ des éléments dominants. Ou bien on a $Ⓗ_w(\mu,\nu) = \mathcal{V}($ ou bien on a $Ⓗ_w(\mu,\nu) = 0$. On a $Ⓗ_w(\mu,\nu) = 0$ si et seulement s'il existe $\alpha \in B_p$ tel que $<\mu, \check{\alpha}> = 0$ et $w\alpha > 0$, ou $<\nu, \check{\alpha}> = 0$ et $w^{-1}\alpha > 0$. En particulier $\tau(\mathcal{V}(wp$ est formé des $(\alpha, 0)$ avec $\alpha \in B_p$ telle que $w\alpha > 0$, et des $(0,\alpha)$ avec $\alpha \in B_p$ telle que $w^{-1}\alpha > 0$.

Ce lemme est un cas particulier de résultats de [14], [11]. Il peut aussi se déduire des résultats similaires de [3] pour les modules de Verma.

Nous noterons w_p l'élément de plus grande longueur du groupe W_p. Soit S un sous-ensemble de B_p. On pose $S' = -w_p S$. C'est un sous-ensemble de B_p.

<u>Lemme 5</u>. On a $d(\mathcal{V}(w_p w_S p, p)) = \# R - \# R_p \cap \mathbb{Z} S$ et $\theta_{w_p w_S} = c\,\pi_S \otimes \pi_{S'}$, où c est une constante non nulle.

La démonstration du lemme sera donnée plus bas.

III. - b) <u>Familles cohérentes et idéaux primitifs</u>.

Soit I un idéal de $U(\underline{g})$. On peut considérer $U(\underline{g})/I$ comme $\underline{g} \times \underline{g}$-module par la formule $(a,b)u = -{}^t a u - ub$ $(a,b \in \underline{g},\ u \in U(\underline{g})/I$. Si V est un $\underline{g} \times \underline{g}$-U-module, on note $|V|$ son image dans le groupe de Grothendieck. On sait que l'on a $e(V) = e_d(|V|)$ si $d = d(V)$ (cf. [13]). Si I est primitif, $U(\underline{g})/I$ est un $g \times g$ - U - module.

<u>Lemme 6</u>. Soit $w \in W$. On définit Ψ_w comme au II, et on pose $I = \ker L(-w p)$.

(i) Pour tout $\mu \in p + P$ dominant régulier, on a $\Psi_w(-\mu) = L(-w\mu)$ et

$$d(\psi_w(-\mu)) = d(L(-wp)) = \frac{1}{2}\, d(U(\underline{g})/\ker L(-w\mu)).$$

(ii) Soit T la famille cohérente de $\underline{g} \times \underline{g}$ - U - modules définie sur $(p+P)\times(p+P)$ telle que $T(p,p) = |U(\underline{g})/I|$. Pour tout $\mu \in p + P$, dominant régulier, on a $T(\mu,\mu) = |U(\underline{g})/\ker L(-w\mu)|$.

(iii) Il résulte de (ii) qu'il existe un polynôme $\xi \in S(\underline{h})$ tel que $\xi(\mu) = e(U(\underline{g})/\ker L(-w\mu))$ pour tout $\mu \in p + P$ dominant régulier. Le polynôme ξ ne dépend que de I, et l'on a degré ξ = 2 degré ψ_w.

<u>Démonstration</u>.

(i) Voir [3] et [4].

(ii) Soient $\mu, \nu \in \underline{h}^*$ des éléments dominants tels que $\mu - \nu \in P$. Il existe un (et, à un facteur constant près, un seul) opérateur d'entrelacement non nul $B(w,\mu,\nu) : \mathcal{L}(\mu,\nu) \to \mathcal{L}(w\mu, w\nu)$, et les $\underline{g} \times \underline{g}$-modules $U(\underline{g})/\ker L(-w\mu)$ et $\mathcal{L}(\mu,\mu)/\ker B(w,\mu,\mu)$ sont isomorphes (cf. [2] proposition 10). Comme les $|\mathcal{L}(\mu,\nu)|$ avec $\mu, \nu \in p + P$ forment une famille cohérente, il nous suffit de voir qu'il existe une famille cohérente T' telle que $T'(\mu,\nu) = |\ker B(w,\mu,\nu)|$ pour tout $\mu, \nu \in p + P$, dominants et réguliers.

Compte tenu des propriétés des familles cohérentes (cf. [14] par exemple) il suffit de démontrer l'assertion suivante. Soient μ, ν dominants réguliers dans

$p+P$, et soit F le $\underline{g}\times\underline{g}$-modules de dimension finie simple de poids extrémal $(\mu-p,\nu-p)$. Alors les modules $(\ker B\,(w,p,p)\otimes F)_{(\mu,\nu)}$ et $\ker B(w,\mu,\nu)$ sont isomorphes, où l'on a noté $V_{(\mu,\nu)}$ la partie d'un $\underline{g}\times\underline{g}$-modules V correspondant au caractère infinitésimal $\chi_\mu\otimes\chi_\nu$.

Notre assertion résulte de ce que $(\mathcal{L}(p,p)\otimes F)_{(\mu,\nu)}$ et $\mathcal{L}(\mu,\nu)$ sont isomorphes (comme on le voit facilement sur la définition donnée plus haut de $\mathcal{L}(\mu,\nu)$) et de ce que $B(w,p,p)\otimes Id_F$ induit dans $\mathcal{L}(\mu,\nu)$ un opérateur d'entrelacement, nécessairement proportionnel à $B(w,\mu,\nu)$. Il faut vérifier que la constante de proportionnalité est non nulle. Ceci résulte par exemple de ce que $(\mathcal{L}\,(p,p)/\ker B(w,p,p)\otimes F)_{(\mu,\nu)}$ est non nul, d'après [14].

(iii) D'après [6], il existe des constantes $A, B > 0$ telle que

$$A\, e(L(\mu))^2 \leq e(U(\underline{g})/\ker L(\mu)) \leq B\, e(L(\mu))^2$$

pour tout $\mu\in\underline{h}^*$. On voit donc que les fonctions polynomiales ψ_w^2 et ξ ont même degré.

<u>Lemme 7</u>. Soit Ⓗ une famille cohérente de $\underline{g}\times\underline{g}$ -modules définie sur $p+P$. On suppose que X = Ⓗ(p) est simple, et on pose $I = \ker X$.

(i) Pour tout $\mu\in p+P$ dominant régulier, on a

$$d(X) = d(Ⓗ(\mu)) = \frac{1}{2}\, d(U(\underline{g})/\ker Ⓗ(\mu)).$$

(ii) Soit T la famille cohérente $\underline{g}\times\underline{g}-U-$modules définie sur $(p+P)\times(p+P)$ telle que $T(p,p) = |U(\underline{g})/\check{I}|$. Pour tout $\mu\in p+P$ dominant régulier, on a $T(\mu,\mu) = |U(\underline{g})/(\ker Ⓗ(\mu))^{\vee}|$.

(iii) Il existe un polynôme $\xi\in S(\underline{h})$ tel que $\xi(\mu) = e(U(\underline{g})/\ker Ⓗ(\mu))$ pour tout $\mu\in p+P$ dominant régulier. Le polynôme ξ ne dépend que de I et l'on a degré ξ = 2 degré ψ.

<u>**Démonstration**</u>. Soit $w\in W$ tel qu'il existe une dualité non dégénérée entre X et $L(-wp)$. Cela existe d'après un résultat de Casselman (cf. [13] lemme 4.5). Soit $\mu\in p+P$ dominant régulier. Il existe encore une dualité non dégénérée entre Ⓗ(μ) et $L(-w\mu)$. En effet, soit F le $\underline{g}$-module simple de dimension finie de poids extrémal $\mu-p$. Il y a une dualité non dégénérée entre Ⓗ$(wp)\otimes F$ et $L(-wp)\otimes F^*$. En gardant les parties de ces modules correspondant aux caractè infinitésimaux χ_μ et $\chi_{-\mu}$, on obtient notre assertion.

Il existe des constantes $A,B > 0$ telles que, si V est un $\underline{g}-K$-module simple, et M un module simple dans la catégorie $\underline{O}$, en dualité non dégénérée, o ait $d(V) = d(M)$ et $A\,e(M)\leq e(V)\leq B\,e(M)$. La démonstration est celle du lemme 3 de [13], compte tenu de ce que les constantes qui y figurent ne dépendent que de $\underline{g}$.

Le lemme 7 résulte maintenant facilement du lemme 6.

Proposition 1.

(i) Soient w et I comme dans le lemme 6. On suppose qu'il existe un sous-ensemble S de B_p tel que $I = \ker L(-w_S p)$. On a alors $d(L(-w p)) = \# R^+ - \# R_p^+ \cap \mathbb{Z} S$ et $\psi_w = c\pi_S$, où c est une constante non nulle.

(ii) Soient $\mathcal{H}$, X et I comme dans le lemme 7. On suppose qu'il existe un sous-ensemble S de B_p tel que $I = \ker L(-w_S p)$. On a alors $d(X) = \# R^+ - \# R_p^+ \cap \mathbb{Z} S$ et $C = c\pi_S$, où c est une constante non nulle.

Démonstration.

(i) Il résulte des lemmes 1 et 6 que l'on a $d(L(-w p)) = \# R^+ - \# R_p^+ \cap \mathbb{Z} S$ et degré ψ_w = degré π_S. on a d'autre part $\tau(I) = S$ (cf. [1]) et il résulte de [13] que ψ_w est divisible par π_S, ce qui termine la démonstration.

(ii) La démonstration de (ii) est analogue.

Démonstration de l'exemple 3 du § I. Il existe un seul idéal primitif de caractère infinitésimal χ_{-p} tel que $\tau(I) = B_p$: c'est l'idéal $\ker L(-w_p p)$ (cf. [1]). On applique la proposition 1.

Démonstration de l'exemple 4 du § I. Il existe un seul idéal primitif de caractère infinitésimal χ_{-p} tel que $\tau(I) = \{\alpha\}$: c'est l'idéal $\ker L(-s_\alpha p)$ (cf. [1]). On applique la proposition 1.

Démonstration du lemme 5. Il résulte de [5] que l'on a $(\ker \mathcal{V}(w_p w_S p, p)) = \ker L(-w_S p) \otimes L(-w_{S'} p)$. On applique la proposition 1.

III. - c) Représentations de W_p.

Les notations sont celles de III-a). Si $d \in \mathbb{N}$, on note ${}^dW_p$ le sous-ensemble de W_p formé des w tels que $d(\mathcal{V}(wp, p)) = d$.

Lemme 8. Soit $d \in \mathbb{N}$. Le $\mathbb{Z}$-module $\sum_{w \in {}^dW_p} \mathbb{Z}\theta_w$ de $S(\underline{h}) \otimes S(\underline{h})$ est stable sous l'action du groupe $W_p \times W_p$.

Démonstration. Soient $w,\sigma,\sigma' \in W_p$. Pour tout $\tau \in W_p$, il existe un entier $n(w,\sigma,\sigma',\tau)$ tel que l'on ait, pour tout $\mu,\nu \in p+P$:

(4) $$\Theta_w(\sigma^{-1}\mu,\sigma'^{-1}\nu) = \sum_{\tau \in W_p} n(w,\sigma,\sigma',\tau)\, \Theta_\tau(\mu,\nu).$$

Pour démontrer (4), on considère pour tout $w' \in W_p$ la famille cohérente $\Xi_{w'}$ telle que $\Xi_{w'}(p,p) = |\mathcal{L}(w'p,p)|$. Les Θ_w sont combinaison linéaire à coefficients dans $\mathbb{Z}$ des $\Xi_{w'}$, et réciproquement. Cela découle des résultat d'Hiraï [8] (cf.[2]) sur la suite de Jordan-Hölder de $\mathcal{L}(wp,p)$. Pour démontre (4), il suffit donc de démontrer l'assertion analogue pour les $\Xi_{w'}$. Dans ce cas, on voit facilement que l'on a $\Xi_{w'}(\sigma^{-1}\mu,\sigma'^{-1}\nu) = \Xi_{\sigma'w'\sigma^{-1}}(\mu,\nu)$.

Supposons de plus $w \in {}^dW_p$. Tous les composants simples de $\Theta_w(\sigma^{-1}p,$ sont de dimension de Gelfand-Kirillov $\leq d$ (car isomorphes à des sous-quotients du produit tensoriel de $\Theta_w(p,p)$ avec une représentation de dimension finie de $\underline{g}\times\underline{g}$, cf.[13]). Il en résulte que l'on a :

(5) $$(\sigma\times\sigma')\theta_w = \sum_{\tau \in {}^dW_p} n(w,\sigma,\sigma',\tau)\theta_\tau.$$

Le lemme 8 rend naturel d'introduire, comme dans Joseph [6], les W_p-modules considérés par Mac Donald dans [9]. Soit S un sous-ensemble de B_p On note V_S le sous-espace $\sum_{w\in W_p} \mathbb{C}\, w\pi_S$ de $S(\underline{h})$. C'est un W_p-module simple. S T un sous-ensemble de B_p. Les modules V_S et V_T sont équivalents si et seulement s'ils sont égaux (cf. [9]).

Il résulte des lemmes 5 et 8 que l'on a

(6) $$\sum_{w\in {}^dW_p} \mathbb{C}\theta_w \supset \bigoplus_d V_S \otimes V_{S'},$$

où $\bigoplus_d$ désigne la somme sur les sous-espaces de la forme V_S contenus dans le sous-espace de $S(\underline{h})$ formé des polynômes homogènes de degré $\frac{1}{2}(\# R - d)$. Il résulte de (6) que l'on a

(7) $$\sum_{w\in W_p} \mathbb{C}\theta_w \supset \bigoplus V_S \otimes V_{S'},$$

où la somme est prise sur l'ensemble des V_S contenus dans $S(\underline{h})$.

Théorème 1. On suppose que le système de racines R_p est produit de systèmes de type A_m. Soit $w \in W_p$. Le polynôme θ_w est homogène et $W_p \times W_p$-harmonique. On a $d(\mathcal{V}(wp,p)) + \text{degré}\,\theta_w = \# R$. De plus les θ_w, où w parcourt W_p, sont linéairement indépendants, et dans les formules (6) et (7) il y a égalité.

<u>Démonstration</u>. Le groupe W_p est un produit de groupes symétriques. Les modules V_S forment un système de représentants de l'ensemble des classes de W_p-modules simples (cf. [9]). On a donc :

$$\dim \oplus V_S \otimes V_{S'} = \Sigma (\dim V_S)^2 = \# W_p .$$

Il résulte de (7) que les θ_w sont linéairement indépendants, et que (7) est une égalité. Il en résulte que (6) est une égalité, ce qui démontre le théorème.

<u>Remarque</u> : Nous laissons au lecteur le soin de modeler sur la conjecture de Jantzen des conjectures sur les θ_w (cf. [7] en particulier).

+
+ +
+

Bibliographie.

1. W. Borho et J.C. Jantzen. Über primitive Ideale in der Einhüllenden einer halbeinfacher Lie-algebra. Inventiones Math. 39(1977) 1-53.

2. M. Duflo. Sur la classification des idéaux primitifs dans l'algèbre enveloppante d'une algèbre de Lie semi-simple. Annals of Math. 105(1977) 107 - 130.

3. J.C. Jantzen. Moduln mit einem höchsten Gewwicht. Habilitationsschrift, Bonn 1977.

4. A. Joseph. Gelfand-Kirillov dimension for the annihilators of simple quotients of Verma modules. J. Lond Math. Soc. A paraître.

5. A. Joseph. On the annihilators of the simple subquotients of the princip series. Ann. Sci. Ec. Norm.Sup., à paraître.

6. A. Joseph. Towards the Jantzen conjecture, I et II. Preprints 1977.

7. A. Joseph. Kostant's problem, Goldie rank and the Gelfand-Kirillov conjecture. Preprint 1978.

8. T. Hiraï. Structure of induced representations and characters of irreducible representations of complex semi-simple Lie groups. Lecture Notes 266 (1972) 167 - 188.

9. I.G. Macdonald. Some irreducible representations of the Weyl groups. Bull. Lond. Math. Soc. 4 (1972) 148 - 150.

10. W. Schmid. Two character identities for semi-simple Lie groups. Lecture Notes 587 (1977) 196 - 225.

11. B. Speh et D. Vogan. Reducibility of generalized principal series representations. A paraître.

12. D. Vogan. Irreducible characters of semi - simple Lie groups I. A paraître

13. D. Vogan. Gelfand-Kirillov dimension for Harish-Chandra modules. A paraître

14. G. Zuckerman. Tensor products of finite and infinite dimensional representations of semi-simple Lie groups. Ann. of Math. 106 (1977) 295-308.

On a fundamental series of
representations related to an affine
symmetric space*

by

Mogens Flensted-Jensen

*This paper contains an exposition given also at the Nancy Strasbourg seminar on "Harmonic analysis on Lie groups", at Kleebach, May 29 - June 2, 1978. An extended version is to appear later.

§1. Introduction.

Let G be a connected non-compact semisimple Lie group and let τ be an involution on G. By H we denote the connected component of the identity of the fixpoint group G^τ. The group G acts by left translation on functions on the (in general non-Riemannian) symmetric space G/H, f.ex. on $L^2(G/H)$ or $C^\infty(G/H)$. We are primarily interested in finding the "discrete series" for G/H, i.e. the representations of G on minimal closed invariant subspaces of $L^2(G/H)$, if such exists.

Example 1.1. A connected non-compact semisimple Lie group G_1 (with finite center) can be identified with the symmetric space $G_1 \times G_1/d(G_1)$, where $d(G_1)$ is the diagonal subgroup, via the mapping $(x,y)d(G_1) \to xy^{-1}$. Then $L^2(G_1 \times G_1/d(G_1)) \cong L^2(G_1)$ and Harish-Chandra's classification of the discrete series gives us the answer in this case, f.ex. discrete series exist if and only if $\operatorname{rank}(G) = \operatorname{rank}(K)$.

Example 1.2. If G_1 is as in example 1.1 and K_1 is a maximal compact subgroup. Then there is no "discrete series", however there is a fundamental series of representations related to the Riemannian symmetric space G_1/K_1. These representations are described by means of the spherical function φ_λ on G_1/K_1, given by Harish-Chandra's formula

$$(1.1) \qquad \varphi_\lambda(x) = \int_{K_1} e^{\langle i\lambda-\rho, H(xk)\rangle} dk, \quad \lambda \in \mathfrak{a}^*_C.$$

The representation T_λ related to φ_λ can be described as the representation of G_1 on the closed linear span in $C^\infty(G_1/K_1)$ of all left translates of φ_λ.

We define in §3 a finite number of families $\{\psi_\lambda\}$ of functions on G/H given by a formula generalizing (1.1). In the case of example 1.1 some of these functions generate the representations in the discrete series and in the limits of discrete series on G_1, if such exists; if discrete series does not exist, then our functions ψ_λ generate a "fundamental series" of representations of G. Let $D(G/H)$ denote the (left) invariant differential operators on G/H, and let $D_R(G)^K$ be the right invariant differential operators on G commuting with the action of a (suitably chosen) maximal compact subgroup K. ($D_R(G)^K$ clearly acts on $C^\infty(G/H)$). The functions ψ_λ satisfy

(1.2) $\qquad \psi \in C^\infty(G/H)$, ψ is K-finite,

(1.3) $\qquad D\psi = \chi_1(D)\psi$, for each $D \in D(G/H)$,

(1.4) $\qquad D\psi = \chi_2(D)\psi$, for each $D \in D_R(G)^K$,

where χ_1, resp. χ_2, is a complex homomorphism of $D(G/H)$, resp. $D_R(G)^K$, explicitly constructed.

If T_λ denote the representation generated by ψ_λ in $C^\infty(G/H)$, then (1.3) in particular shows that T_λ admits a central character.

We want to think of (1.4) as determining the "minimal K-type" δ_λ of T_λ, as such δ_λ should occur with multiplicity one. (In example 1.1 this is true, at least when T_λ is in the discrete series, since ψ_λ is then essentially the "spherical trace function for T_λ of type δ_λ", where δ_λ is the minimal K-type of T_λ)[1]. We attack this point in §4 by a generalization to G/H of a "weak Blattner conjecture".

In §5 we find a sufficient condition for discrete series on G/H to exist, namely

(1.5) $\quad \operatorname{rank}(G/H) = \operatorname{rank}(K/K \cap H).$

This condition is sufficient since we show that $\psi_\lambda \in L^2(G/H)$ for λ "sufficiently regular".

Our methods are rather simple compared to the deep analysis and algebra involved in the study so far of discrete series for G_1.

Most of our results are obtained by a reduction to function theory on a Riemannian symmetric space G^0/H^0, "dual to G/H". This reduction comes about via a method of analytic continuation introduced in [3].

)[1] see [10].

§2. Structure and duality for semisimple symmetric spaces.

Let $\mathfrak{g}$ be a real semisimple Lie algebra of the non-compact type. Let τ be any involutive automorphism (involution). Then [1] there exists a Cartan involution σ of $\mathfrak{g}$ such that $\sigma\tau = \tau\sigma$. Let $\mathfrak{g} = \mathfrak{h} + \mathfrak{q}$, resp. $\mathfrak{g} = \mathfrak{k} + \mathfrak{p}$, be the decomposition of $\mathfrak{g}$ into +1 and -1 eigenspaces for τ, resp. σ. Since τ and σ commute, $\mathfrak{g}$ decomposes into the direct sum of vectorspaces:

$$\mathfrak{g} = \mathfrak{h} \cap \mathfrak{k} + \mathfrak{q} \cap \mathfrak{k} + \mathfrak{h} \cap \mathfrak{p} + \mathfrak{q} \cap \mathfrak{p}.$$

Let $\mathfrak{g}_0$ be the subalgebra fixed under the involution $\sigma\tau$, i.e. $\mathfrak{g}_0 = \mathfrak{h} \cap \mathfrak{k} + \mathfrak{q} \cap \mathfrak{p}$. $\mathfrak{g}_0$ and $\mathfrak{h}$ are reductive Lie algebras with Cartan-decompositions given by the restriction of σ.

Let $\mathfrak{g}_{\mathbb{C}} = \mathfrak{g} + i\mathfrak{g}$ be the complexification of $\mathfrak{g}$. Extend τ and σ to complex linear involutions of $\mathfrak{g}_{\mathbb{C}}$. Define the real subalgebras $\mathfrak{g}^0$, $\mathfrak{h}^0$ and $\mathfrak{k}^0$ of $\mathfrak{g}_{\mathbb{C}}$, dual to $\mathfrak{g}$, $\mathfrak{h}$ and $\mathfrak{k}$ as follows:

$$\begin{aligned}
\mathfrak{g}^0 &= \mathfrak{g}_0 + i(\mathfrak{q} \cap \mathfrak{k} + \mathfrak{h} \cap \mathfrak{p}) \\
\mathfrak{h}^0 &= \mathfrak{h}_{\mathbb{C}} \cap \mathfrak{g}^0 = \mathfrak{h} \cap \mathfrak{k} + i(\mathfrak{h} \cap \mathfrak{p}) \\
\mathfrak{k}^0 &= \mathfrak{k}_{\mathbb{C}} \cap \mathfrak{g}^0 = \mathfrak{h} \cap \mathfrak{k} + i(\mathfrak{q} \cap \mathfrak{k}).
\end{aligned}$$

Let σ^0 and τ^0 (or just σ and τ) denote the restrictions of σ and τ to $\mathfrak{g}^0$. Now notice that $(\mathfrak{g}^0, \sigma^0, \tau^0, \mathfrak{h}^0, \mathfrak{k}^0)$ satisfies the same conditions as above $(\mathfrak{g}, \tau, \sigma, \mathfrak{k}, \mathfrak{h})$. So in particular τ^0 is a Cartan involution, and $\mathfrak{h}^0$ is a maximal compact subalgebra of $\mathfrak{g}^0$. Notice that $\mathfrak{g}_0 = \mathfrak{h} \cap \mathfrak{k} + \mathfrak{p} \cap \mathfrak{q}$ corresponds to itself under the duality.

Let $G_{\mathbb{C}}$ be a simply connected complex linear Lie group with Lie algebra $\mathfrak{g}_{\mathbb{C}}$. Let G, G^0, H, H^0, K, K^0 and G_0 be the analytic subgroups of $G_{\mathbb{C}}$ corresponding to $\mathfrak{g}, \mathfrak{g}^0, \mathfrak{h}, \mathfrak{h}^0, \mathfrak{k}, \mathfrak{k}^0$ and $\mathfrak{g}_0$. Let $\mathfrak{u} = \mathfrak{k} + i\mathfrak{p}$, then $\mathfrak{u}$ is a compact real form of $\mathfrak{g}_{\mathbb{C}}$, and U, the corresponding analytic subgroup, is maxmal compact in $G_{\mathbb{C}}$ and simple connected.

Notice that $G \cap U = K$, $G^0 \cap U = H^0$ and that $L = H \cap K = H^0 \cap K^0$ is connected. The fibration theory of symmetric spaces (see f.ex. Loos [8]) asserts that G/H is isomorphic to $K \times_L G_0/L$, i.e. the mapping $(k,X) \to k \exp X\, H$ defines a diffeomorphism of the quotient $(K \times \mathfrak{p} \cap \mathfrak{q})/\sim$ onto G/H, where $(k,X) \sim (k',X')$ if and only if there exists an $\ell \in L$ such that $k = k'\ell$ and $X = \mathrm{Ad}(\ell^{-1})X'$. Also

(2.1) $\qquad (X,Y,h) \to \exp X \exp Y h$

is a diffeomorphism of $i(\mathfrak{k} \cap \mathfrak{q}) \times \mathfrak{p} \cap \mathfrak{q} \times H^0$ onto G^0.

From this the following lemma is easy:

Lemma 2.1. (i) Assume that $f \in C^\infty(K \times G_0)$ satisfies

$$f(k\ell, \ell^{-1}x\ell_1) = f(k,x) \quad \text{for all } k \in K,\ x \in G_0 \text{ and } \ell, \ell_1 \in L$$

then (by abuse of notation)

$$f(kxH) = f(k,x)$$

is well defined in $C^\infty(G/H)$.

(ii) Assume that $f \in C^\infty(K^0 \times G_0)$ satisfies

$$f(k\ell, \ell^{-1}x\ell_1) = f(k,x)$$

for all $k \in K^0$, $x \in G_0$ and $\ell, \ell_1 \in L$ then (by abuse of notation)

$$f(kxH^0) = f(k,x)$$

is well defined in $C^\infty(G^0/H^0)$.

Let now $G^\sim$ be the simply connected convering group of G. Let $K^\sim$, $H^\sim$, $L^\sim$ and $G_0^\sim$ be the analytic subgroups of G convering K,H,L and G_0. Let $K_{\mathbb{C}}^\sim$ be a simply connected complex Lie group containing $K^\sim$, and having Lie algebra $\mathfrak{k}_{\mathbb{C}}$. (This is possible since $K^\sim$ is isomorphic to a direct product of a vectorgroup and a simply connected compact semisimple Lie group). Let $K^{0\sim}$ be the analytic subgroup of $K_{\mathbb{C}}^\sim$ with Lie algebra $\mathfrak{k}^0$. Then (ii) of the lemma also holds with $K^{0\sim}$, $G_0^\sim$ and $L^\sim$ instead of K^0, G_0 and L. (G^0/H^0, being simply connected, is unchanged, see (2.1)).

Let $D_R(G^\sim)$ and $D_R(G^0)$ be $U(\mathfrak{g}_{\mathbb{C}})$ considered as right invariant differential operators on $G^\sim$ and G^0, $(X\varphi(x) = \frac{d}{dt}\varphi(\exp(-tX)x)_{t=0})$, and let $D \to D^\eta$ be the hereby defined isomorphism of $D_R(G^\sim)$ onto $D_R(G^0)$. Similarly $U(\mathfrak{g}_{\mathbb{C}})^{\mathfrak{h}_{\mathbb{C}}}$ can be used to identify $D(G^\sim/H^\sim)$ and $D(G^0/H^0)$, (the left invariant differential operators $(X\varphi)(x) = \frac{d}{dt}\varphi(x\ \exp tX)_{t=0})$.

The following theorem is now a rather simple consequence of the lemma. It generalizes Theorem 4.2 and Theorem 4.3 in [3].

<u>Theorem 2.2.</u> Let $C^\infty_{K^\sim}(G^\sim/H^\sim)$ denote the $K^\sim$-finite functions in $C^\infty(G^\sim/H^\sim)$, and $C^\infty_{K^0}(G^0/H^0)$ the K^0-finite functions on G^0/H^0.

(i) There is a bijection $f \to f^\eta$ of $C^\infty_{K^\sim}(G^\sim/H^\sim)$ onto $C^\infty_{K^0}(G^0/H^0)$, satisfying $f_{|G_0} = f^\eta{}_{|G_0}$ and for each $x \in G_0$ $k \to f(kx)$ extends to a holomorphic function of $K^\sim_{\mathbb{C}}$ s.t. $f^\eta(kx)$ is the restriction of this function to $K^{0\sim}$.

(ii) For any $D \in D_R(G^\sim) \cup D(G^\sim/H^\sim)$

$$(Df)^\eta = D^\eta f^\eta. \qquad f \in C^\infty_{K^\sim}(G^\sim/H^\sim)$$

(i.e. $f \to f^\eta$ is a $U(\mathfrak{g}_{\mathbb{C}})$ left module isomorphism and a $U(\mathfrak{g}_{\mathbb{C}})^{\mathfrak{h}_{\mathbb{C}}}$ right module isomorphism.)

From here on we shall just write $f = f^\eta$ and $D = D^\eta$.

§3. An integral formula.

We start this section with some heuristic arguments:

Assume $E \subset C^\infty(G^\sim/H^\sim)$ is an $G^\sim$ invariant irreducible subspace, and $\varphi \in E$ is $K^\sim$-finite of type δ, $\delta \in (K^\sim)^\wedge$. Then φ should be an eigenfunction for all $D \in D(G^\sim/H^\sim)$. If furthermore δ occurs in E with multiplicity one then φ should be an eigenfunction of each $D \in D_R(G^\sim)^{K^\sim}$. That is φ should satisfy (1.2), (1.3) and (1.4).

In the following we are going to describe a family of functions ψ_λ, λ in a certain index set satisfying (1.2) and (1.3), such that ψ_λ is of a specific $K^\sim$-type δ_λ, $\delta_\lambda \in (K^\sim)^\wedge$. We also describe χ_1^λ, corresponding to ψ_λ. We shall see in §4 that δ_λ is contained with multiplicity one in E^λ, the $G^\sim$ invariant subspace generated by ψ_λ ; and that ψ_λ satisfies (1.4).

Since we look for joint eigenfunctions of $D(G^0/H^0)$ on the Riemannian symmetric space G^0/H^0, it is natural to look for these functions in terms of their Poisson-integral formula.

Let $\mathfrak{t}$ be a maximal Abelian subalgebra of $i(\mathfrak{k}\cap\mathfrak{q})$ extend $\mathfrak{t}$ to a maximal Abelian subalgebra $\mathfrak{a}$ of $\mathfrak{q}^0 = i(\mathfrak{k}\cap\mathfrak{q}) + \mathfrak{p}\cap\mathfrak{q}$, such that $\mathfrak{a} = \mathfrak{t} + \mathfrak{a}_1$, where $\mathfrak{a}_1 \subset \mathfrak{p}\cap\mathfrak{q}$. (All such choices are conjugate under $L = K \cap H$). Let $\Delta_{\mathfrak{t}} \subset \mathfrak{t}^*$ be the system of roots of in $\mathfrak{k}^0$, and let $W_{\mathfrak{t}}$ be the corresponding Weyl group. Choose a positive Weylchamber $\mathfrak{t}^+$ in $\mathfrak{t}$. Let $\Delta_{\mathfrak{t}}^+$ be the set of positive roots and let $\rho_{\mathfrak{t}}$ be half the sum of the positive roots counted with multiplicity.

Notice that K^0/L ($\simeq K^{0\sim}/L^\sim$) is a Riemannian symmetric space, which is non-compact. The spherical functions on K^0/L are given by

$$(3.1) \qquad \varphi_\nu(y) = \int_L e^{\langle i\nu-\rho_t, H(yk)\rangle} dk, \quad \nu \in \mathfrak{t}^*_{\mathbb{C}}.$$

By [5] , the finite dimensional irreducible representations δ of K^0 with an L-fixed vector are determined by the set

$$(3.2) \qquad M^+ = \{\mu \in \mathfrak{t}^*_{\mathbb{C}} \mid \frac{\langle\mu,\beta\rangle}{\langle\beta,\beta\rangle} \in \mathbb{Z}^+ \text{ for each } \beta \in \Delta^+_{\mathfrak{t}}\}.$$

If $\mu \in M^+$ then μ is the dominant weight of the corresponding δ_μ, (to be precise μ should be extended to be zero on the compact part of a Cartan subalgebra of $\mathfrak{k}^0$ containing $\mathfrak{t}$), and the corresponding spherical function is φ_ν, with $i\nu = \mu + \rho_{\mathfrak{t}}$.

If we let e_μ be a dominant weight vector and e_L the L-fixed vector for δ_μ, then by choosing a suitable scalar product:

$$(3.3) \qquad e^{\langle i\nu-\rho_{\mathfrak{t}}, H(y)\rangle} = (\delta(y)e_\mu, e_L) \ , \quad y \in K^0.$$

Thus $e^{\langle i\nu-\rho_{\mathfrak{t}}, H(yk)\rangle}$ is a K^0-finite function of type $\check{\delta}_\mu$ [1)] for each $k \in L$, and extends to a holomorphic function on $K^\sim_{\mathbb{C}}$. Hereby (3.2) also parametrizes the finite-dimensional irreducible representations of $K^\sim$ having an $L^\sim$-fixed vector. As a representation of $K^\sim$ δ_μ is unitary if and only if

---

1) $\check{\delta}$ is the contragridient representation to δ .

$\mu \in M^+ \cap \mathfrak{t}^*$.

Let now $\Delta \subset \mathfrak{a}^*$ be the system of roots of $\mathfrak{a}$ in $\mathfrak{g}^0$, and W the corresponding Weyl group. Choose a positive Weyl chamber $\mathfrak{a}^+$ in $\mathfrak{a}$, with $\Delta^+ = \Delta^+(\mathfrak{a}^+)$ the corresponding system of positive roots, such that

$$\beta \in \Delta,\ \beta_{|\mathfrak{t}} \in \Delta^+_{\mathfrak{t}} \Rightarrow \beta \in \Delta^+. \tag{3.4}$$

Let $\rho = \rho(\mathfrak{a}^+)$ be half the sum of the positive roots counted with multiplicity. Let $\mathfrak{g}^0 = \mathfrak{h}^0 + \mathfrak{a} + \mathfrak{n}$ and $G^0 = H^0AN$ be the related Iwasawa decompositions. For $x \in G^0$ define $H(x) \in \mathfrak{a}$ such that

$$x \in H^0 \exp H(x) N.$$

Now for each $\lambda \in \mathfrak{a}^*_{\mathbb{C}}$ define

$$\psi_\lambda(x) = \int_L e^{\langle -i\lambda-\rho, H(x^{-1}k)\rangle}\, dk, \quad x \in G^0. \tag{3.5}$$

By well known facts about symmetric spaces $\psi_\lambda \in C^\infty(G^0/H^0)$ and

$$D\psi_\lambda = \chi_\lambda(D)\psi_\lambda \quad \text{for each} \quad D \in D(G^0/H^0),$$

where $\chi_\lambda(D)$ is the eigenvalue for D of the spherical function

$$\Phi_\lambda(x) = \int_{H^0} e^{\langle -i\lambda-\rho, H(x^{-1}k)\rangle}\, dk = \int_{H^0} e^{\langle i\lambda-\rho, H(xk)\rangle}\, dk, \quad x \in G^0$$

Since the Jacobian of the action of $y \in K^0$ on L is $e^{\langle 2\rho_{\mathfrak{t}}, H(y^{-1}k)\rangle}$ we get by standard technics that

(3.6) $$\psi_\lambda(yx) = \int_L e^{\langle i\lambda+\rho-2\rho_t, H(yk)\rangle} e^{\langle -i\lambda-\rho, H(x^{-1}k)\rangle} dk$$

for all $y \in K^0$ and $x \in G^0$.

If we now assume, that λ belongs to the set

(3.7) $$\Lambda = \Lambda(\mathfrak{a}^+) = \{\lambda \in \mathfrak{a}^*_{\mathbb{C}} \mid \mu_\lambda = (i\lambda+\rho-2\rho_{\mathfrak{t}})|_{\mathfrak{t}} \in M^+\}$$

then ψ_λ is K^0-finite of type $\check{\delta}_{\mu_\lambda}$.

Summing up we have

<u>Theorem 3.1</u>. For each positive Weylchamber $\mathfrak{a}^+$ satisfying (3.4), define

$$\psi_\lambda(x) = \int_L e^{\langle -i\lambda-\rho, H(x^{-1}k)\rangle} dk, \quad x \in G^0,$$

$\lambda \in \Lambda(\mathfrak{a}^+)$, then

(i) $\psi_\lambda \in C^\infty_{K^0}(G^0/H^0)$ is K^0-finite of type $\check{\delta}_{\mu_\lambda}$, $\mu_\lambda = i\lambda+\rho-2\rho_{\mathfrak{t}} |_{\mathfrak{t}}$.

(ii) $D\psi_\lambda = \chi_\lambda(D)\psi_\lambda$ for each $D \in D(G^0/H^0)$,

where $\chi_\lambda(D)$ is the eigenvalue for the spherical function Φ_λ.

<u>Remark 3.2</u>. Transporting ψ_λ to $G^\sim/H^\sim$ we see that ψ_λ has the desired properties, except condition (1.4), which we shall look at in §4.

<u>Remark 3.3</u>. Clearly ψ_λ depends on the choices of $\mathfrak{t}^+$ and $\mathfrak{a}^+$ satisfying (3.4). The essentially different choises can be parametrized by the closed orbits of K^0 on G^0/MAN, where M is the centralizer of A in H^0.

These orbits are in one-to-one correspondance with the cosetspace W_σ/W_L, where $W_\sigma = \{w \in W \mid w(\mathfrak{t}) = \mathfrak{t}\}$ and $W_L = \{w \in W \mid w = Ad(m'),\ m' \in L\}$, (see [9]).

§4. The "weak Blattner conjecture".

We now fix a positive Weyl chamber $\mathfrak{a}^+$ with the property (3.4). Let $\lambda \in \Lambda = \Lambda(\mathfrak{a}^+)$, and let E^λ be the closed subspace of $C^\infty(G^\sim/H^\sim)$ generated by ψ_λ, and T_λ the representation of $G^\sim$ on E^λ. Let $E^\lambda_{K^\sim} = \sum_{\delta \in K^{\sim\wedge}} E^\lambda(\delta)$ be the $K^\sim$-finite functions in E^λ. We denote by $m_\lambda(\delta) = \frac{\dim E^\lambda(\delta)}{\dim \delta}$ the multiplicity of δ in E^λ. Since ψ_λ is analytic and $K^\sim$-finite and $G^\sim$ is connected, it follows that $D_R(G^\sim)\psi_\lambda = E^\lambda_{K^\sim}$.

We shall now look for those $\check{\delta}\mu$, $\mu \in M^+$, (see (3.2)), which occur in E^λ. Since every $0 \neq f \in C^\infty(G^\sim/H^\sim)$ can be translated to a function which is non-zero at $eH^\sim$, it follows that every $G^\sim$-invariant subspace of $C^\infty(G^\sim/H^\sim)$ contain such a δ-type (i.e. has an $L^\sim = K^\sim \cap H^\sim$ fixed vector).

Let $\Delta_2^+ = \{\alpha \in \Delta^+ \mid \alpha_{|\mathfrak{t}} = 0\}$ and $\Delta_1^+ = \Delta^+ \setminus \Delta_2^+$. Define complex nilpotent subalgebras of $\mathfrak{g}_{\mathbb{C}}$ by $\mathfrak{n}_{\mathfrak{t}} = \sum_{\beta \in \Delta_{\mathfrak{t}}^+} \mathfrak{g}_\beta$

$$\mathfrak{n}_1 = \sum_{\alpha \in \Delta_1^+} \mathfrak{g}_\alpha\ , \quad \mathfrak{n}_2 = \sum_{\alpha \in \Delta_2^+} \mathfrak{g}_\alpha\ ,$$

and similarly $\bar{\mathfrak{n}}_{\mathfrak{t}} = \sum_{\beta \in \Delta^+} \mathfrak{g}_{-\beta}$, etc. Notice that $\mathfrak{n}_{\mathfrak{t}} \subset \mathfrak{n}_1$,

$\sigma(\mathfrak{n}_1) = \mathfrak{n}_1$, $\mathfrak{n}_1 + \mathfrak{n}_2 = \mathfrak{n}_{\mathbb{C}}$ and $\sigma(\mathfrak{n}_2) = \overline{\mathfrak{n}}_2 \subset \mathfrak{l}_{\mathbb{C}} + \mathfrak{n}_{\mathbb{C}}$.

Let $N = \dim \mathfrak{n}_1/\mathfrak{n}_t$. Choose $\gamma_1 \leq \cdots \leq \gamma_N \in \Delta_1^+$ and $X_i \in \mathfrak{g}_{-\gamma_i}$, $i = 1,\cdots,N$ such that $X_1,\cdots,X_N$ form a basis for a $M \cap L = M_0$-invariant complement of $\overline{\mathfrak{n}}_t$ in $\overline{\mathfrak{n}}_1$. We shall call the γ_i's for "non-compact roots".

We recall that $\lambda \in \mathfrak{a}_{\mathbb{C}}^*$ is called simple if the Poisson-transform

$$(4.1) \quad S \to \int_{H^0/M} e^{<-i\lambda-\rho, H(x^{-1}hM)>} S(hM)\,dhM = P_\lambda(S,x)$$

of the set of distributions on H^0/M into $C^\infty(G^0/H^0)$ is injective. It is known, see [6], that λ is simple if $\mathrm{Im}\lambda \in \overline{\mathfrak{a}^{*+}}$. Define

$$\Lambda_0 = \{\lambda \in \Lambda \mid \mathrm{Im}\lambda \in \overline{\mathfrak{a}^{*+}}\}.$$

<u>Theorem 4.1</u>. (The "weak Blattner conjecture").

Let $\lambda \in \Lambda_0$ and $\mu \in M^+$ then

$$(4.2) \quad m_\lambda(\mu) = m_\lambda(\delta_\mu^\vee) < +\infty.$$

and in particular

$$(4.3) \quad m_\lambda(\mu_\lambda) = 1$$

(4.4) If $m_\lambda(\mu) \neq 0$, then there exist $w \in W_t$ and $n_1,\cdots n_N \in \mathbb{Z}^+$ s.t. $w(\mu+\rho_t) = \mu_\lambda + \rho_t + \sum_{i=1}^{N} n_i\gamma_i$.

<u>Outline of proof:</u> $\mathfrak{g}_{\mathbb{C}} = \ell_{\mathbb{C}} + \sum_{i=1}^{N} \mathbb{C}X_i + \mathfrak{a}_{\mathbb{C}} + \mathfrak{m}_{\mathbb{C}} + \mathfrak{n}_{\mathbb{C}}$ and

$$\ell_{\mathbb{C}} \cap \left(\sum_{i=1}^{N} \mathbb{C} X_i + \mathfrak{a}_{\mathbb{C}} + \mathfrak{m}_{\mathbb{C}} + \mathfrak{n}_{\mathbb{C}} \right) = \mathfrak{m}_{\mathbb{C}} \cap \ell_{\mathbb{C}} = \mathfrak{m}_{0\mathbb{C}}$$

Therefore for every $D \in U(\mathfrak{g}_{\mathbb{C}})$ there exists

$$D'' = \sum_{j=1}^{N_1} X^{\nu^j} D_j \ , \tag{4.5}$$

where $X^{\nu} = X_1^{\nu_1} \cdots X_N^{\nu_N}$, $\nu = (\nu_1, \cdots, \nu_N)$ and $D_j \in U(\mathfrak{a}_{\mathbb{C}}) = S(\mathfrak{a}_{\mathbb{C}})$, such that

$$D - D'' \in \ell_{\mathbb{C}} U(\mathfrak{g}_{\mathbb{C}}) + U(\mathfrak{g}_{\mathbb{C}})(\mathfrak{m}_{\mathbb{C}} + \mathfrak{n}_{\mathbb{C}}).$$

Let $|\nu^j| = \sum_{i=1}^{N} \nu_i{}^j \gamma_i$. For $\varphi \in C^{\infty}(H^0/M)$ we define $\varphi^{\sim} \in C^{\infty}(G^0)$ by $\varphi^{\sim}(han) = \varphi(h) e^{\langle -i\lambda-\rho, H(a)\rangle}$, for $h \in H^0$, $a \in A$, $n \in N$. For $D \in U(\mathfrak{g}_{\mathbb{C}})$ we define a distribution T_D on H^0/M by

$$T_D(\varphi) = \int_L (D\varphi^{\sim})(k)\,dk$$

(where D acts as a left invariant differential operator).

We can now take φ to be a $L^{\sim}$-fixed vector in $E^{\lambda}(\check{\delta}_{\mu})$, and $D \in D_R(G^{\sim})^{L^{\sim}}$, such that $D\psi_{\lambda} = \varphi$. By (4.5), (4.1) and the formula for ψ_{λ} we get

$$D\psi_{\lambda} = \sum_{j=1}^{N_1} \langle D_j, -i\lambda-\rho \rangle P_{\lambda}(T_{X^{\nu^j}}, x) = \varphi. \tag{4.6}$$

$\varphi \in E^{\lambda}(\check{\delta}_{\mu})$ and therefore φ is an eigenfunction of $U(\mathfrak{h}_{\mathbb{C}})^{\ell_{\mathbb{C}}}$ $(\subset D_R(G^{\sim})^{L^{\sim}})$. For $Z \in U(\mathfrak{h}_{\mathbb{C}})^{\ell_{\mathbb{C}}}$, let $Z = Z' + Z'' + Z'''$ according to

$$U(\mathfrak{h}_{\mathbb{C}}) = \ell_{\mathbb{C}} U(\mathfrak{h}_{\mathbb{C}}) + U(\mathfrak{t}_{\mathbb{C}}) + U(\mathfrak{t}_{\mathbb{C}} + \mathfrak{n}_{t})\mathfrak{n}_{t}.$$

Then we get

$$(4.7)\quad ZD\psi_\lambda = \sum_{j=1}^{N_1} <Z'',\varepsilon_j><D_j,-i\lambda-\rho>P_\lambda(T_{X^{\nu^j}},x) + \sum_{j=1}^{N_1} <D_j,-i\lambda-\rho>P_\lambda(T_{Z'''X^{\nu^j}},x),$$

where $\varepsilon_j = (-i\lambda-\rho-|\nu^j|)_{|\mathfrak{t}}$,

Now we know, since φ is of type $\check{\delta}_\mu$, that $Z\varphi = \chi_3(Z)\varphi$ where

$$(4.8)\quad \chi_3(Z) = <Z'',-\mu-2\rho_{\mathfrak{t}}> = <Z'',w(-\mu-\rho_{\mathfrak{t}})-\rho_{\mathfrak{t}}>,\ w \in W_{\mathfrak{t}}$$

and that $\mu+\rho_{\mathfrak{t}}$ is determined up to $W_{\mathfrak{t}}$-conjugacy by χ_3. Now using that λ is simple one can show from (4.6), (4.7) and (4.8) that there is a $w \in W_t$ and a j such that

$$w(\mu+\rho_{\mathfrak{t}}) = \mu_\lambda + \rho_{\mathfrak{t}} + |\nu_j|.$$

(4.2) and (4.3) are now consequences of (4.4) and (4.6), and the theorem follows

Q.e.d.

<u>Corollary 4.4</u>. (of the proof).

(i) For $D \in D_R(G^\sim)^{K^\sim}$ we can find a $D'' \in U(\mathfrak{a}_{\mathbb{C}}) = S(\mathfrak{a}_{\mathbb{C}})$ such that $D - D'' \in \ell_{\mathbb{C}}\, U(\mathfrak{g}_{\mathbb{C}}) + U(\mathfrak{g}_{\mathbb{C}})(\mathfrak{m}_{\mathbb{C}}+\mathfrak{n}_{\mathbb{C}})$.

(ii) For each $\lambda \in \mathfrak{a}^*_{\mathbb{C}}$ ψ_λ is an eigenfunction of all $D \in D_R(G^0)^{K^0}$ with the eigenvalue

$$\chi'_\lambda(D) = <D'',-i\lambda-\rho>.$$

<u>Proof</u>: Since $D \in D_R(G^\sim)^{K^\sim}$, $D\psi_\lambda$ is of type $\check{\delta}\mu_\lambda$, for each $\lambda \in \Lambda_0$, so (i) follows from (4.6) and (4.7). (ii) follows from (i).

Q.e.d.

Corollary 4.5. For each $\lambda \in \Lambda_0$

(i) T_λ is indecomposable.

(ii) If T_λ is unitarizable, then T_λ is irreducible.

(iii) In particular if $|\psi_\lambda| \in L^2(G/H)$, then T_λ is irreducible.

§5. Existence of discrete series.

In this section we indicate some properties of ψ_λ which can be derived from the formula (3.6). To ensure that $\delta\mu_\lambda$ is unitary (as a representation of $K^\sim$) we require $\lambda_{|\mathfrak{t}} \in i\mathfrak{t}^*$. Then $|\psi_\lambda(zx)| = |\psi_\lambda(x)|$ for $x \in G^\sim$, $z \in \text{centre}(G^\sim)$, thus $|\psi_\lambda|$ is well defined on G/H. We shall say that T_λ is "discrete" if

$$|\psi_\lambda| \in L^2(G/H).$$

Conjecture 5.1. There exists discrete series on $G^\sim/H^\sim$ if and only if $\text{rank}(G/H) = \text{rank}(K/L)$. (i.e. if and only if $\mathfrak{a} = \mathfrak{t} \subset i(\mathfrak{k} \cap \mathfrak{q})$).

Theorem 5.2. Let $\operatorname{rank}(G/H) = \operatorname{rank}(K/L)$. Then

(i) $$|\psi_\lambda(x)| \le 1, \quad \text{for} \quad x \in G_C,$$

if $\lambda \in \Lambda \cap it^*$ and $\langle i\lambda+\rho,\beta\rangle \ge 0$ for all $\beta \in \Delta^+$.

(ii) There exists a constant $\zeta > 0$ such that

$$|\psi_\lambda| \in L^2(G/H)$$

whenever $\lambda \in \Lambda \cap it^*$ and $\langle i\lambda,\beta\rangle > \zeta$ for all $\beta \in \Delta^+$.

Outline of proof:

Let $\nu_1,\cdots,\nu_r \in i\mathfrak{a}^*$ be the "fundamental weights", and $\pi_1,\cdots,\pi_r$ finite dimensional representations of $G_{\mathbb{C}}$ and $\mathfrak{g}_{\mathbb{C}}$ on $V_1,\cdots,V_r$ with dominant weights $\nu_1,\cdots,\nu_r$ (cf. (3.2)). For each $j = 1,\cdots,r$ choose a norm on V_j such that π_j is unitary on U, and choose unit weight vectors ξ_j for V_j belonging to ν_j. Then similar to [4], page 294, with $\lambda = \sum_{j=1}^{r} \lambda_j\nu_j$ and $\rho = \sum_{j=1}^{r} \rho_j\nu_j$,

$$e^{\langle -i\lambda-\rho,H(x)\rangle} = \prod_{j=1}^{r} \|\pi_j(x)\xi_j\|^{(-i\lambda_j-\rho_j)}, \quad x \in G^0$$

and

$$\psi_\lambda(x) = \int_L \prod_{j=1}^{r} \|\pi_j(x^{-1})\pi_j(k)\xi_j\|^{-i\lambda_j-\rho_j} dk. \tag{5.2}$$

It is easily seen that $H(\sigma(x)) = H(x)$[1], so if $x \in \exp(\mathfrak{p}\cap\mathfrak{q})$, and $k \in L$, then $H(x^{-1}k) = H(xk)$.

---

[1] Since $\mathfrak{a} = \mathfrak{t} \subset \mathfrak{h}_{\mathbb{C}}$

Now choose $\mathfrak{b}$ maximal Abelian i $\mathfrak{p} \cap \mathfrak{q}$, diagonalize $\pi_j(\mathfrak{b})$, $j = 1,\cdots,r$ by choosing an orthonormal basis of weight vectors $v_{1j},\ldots,v_{d_j,j}$ in V_j. Let v_{ij} have weight $\mu_{ij} \in \mathfrak{b}^*$.

Let now $k \in L$ and $\pi_j(k)\xi_j = \sum_{i=1}^{d_j} b_{ij} v_{ij}$ then for $H \in \mathfrak{b}$:

$$\|\pi_j(\exp H)\pi_j(k)\xi_j\|^2 = \sum_{i=1}^{d_j} e^{2<\mu_{ij},H>} |b_{ij}^2|$$

which is also equal to

$$\|\pi_j(\exp(-H))\pi_j(k)\xi_j\|^2 = \sum_{i=1}^{d_j} e^{-2<\mu_{ij},H>} |b_{ij}^2|$$

thus

$$(5.3) \quad \|\pi_j(\exp H)\pi_j(k)\xi_j\| = \sum_{i=1}^{d_j} \cosh(2<\mu_{ij},H>)|b_{ij}^2| \geq \|\xi_j\|^2 = 1$$

From which (i) follows. Elaborating on (5.3) we get the result (ii).

Remark 5.3. Even in the case of example 1.1 we feel that our approach may be of interest. We should mention that in this case $G = G_1 \times G_1$, $H = d(G_1)$, $K = K_1 \times K_1$ and $G^0 \approx G_{1\mathbb{C}}$, $H^0 \approx U_1$, $K^0 \approx K_{1\mathbb{C}}$, also $L = d(K_1)$, such that $K^\wedge(L)$ is precisely the set $\{\delta \otimes \check{\delta} \mid \delta \in K_1{}^\wedge\}$.

R E F E R E N C E S

1. Berger, M., Les espaces symétriques non compacts, Ann. Sci. École Norm. Sup. 74 (1957), 85-177.

2. Duflo, M., Representation de carré intégrable des groupes semisimple réels. Sem. Bourbaki, exp. 508, 1977/78.

3. Flensted-Jensen, M., Spherical functions on a real semisimple Lie group. A method of reduction to the complex case. Journ. Func. Anal. (1978), to appear.

4. Harish-Chandra, Spherical functions on a semisimple Lie group, I. Amer. Journ. Math. 80 (1958), 241-310.

5. Helgason, S., A duality for symmetric spaces with applications to group representations. Adv. Math. 5 (1970), 1-154.

6. Helgason, S., A duality for symmetric spaces with applications to group representations, II. Differential equations and eigenspace representations. Adv. Math. 22 (1976), 187-219.

7. Kostant, B., On the existence and irreducibility of certain series of representations, In "Lie groups and their representations" (I.M. Gelfand. Ed.), 231-329, Halsted Press, New York 1975.

8. Loos, O., Symmetric spaces, I: General theory, New York - Amsterdam, W.A. Benjamin, Inc. 1969.

9. Matsuki, T., The orbits of affine symmetric spaces under the action of minimal parabolic subgroups. Preprint, Dept. of Math., Hiroshima University.

10. Vogan, D., Algebraic structure of irreducible representations of semisimple Lie groups. Preprint, Dept. of Math., M.I.T.

HIGHER ORDER TENSOR PRODUCTS OF WAVE EQUATIONS

Hans Plesner Jakobsen*

Introduction. Let T be a unitary (irreducible) representation of a group G in a space H of holomorphic functions on a homogeneous domain $D \subseteq \mathbb{C}^n$. The problem considered here is that of describing the decomposition of $T^n = \underset{n}{\otimes} T$. Of special interest is the situation in which the elements of H are solutions to "wave-equations". This is of relevance to theoretical physics ([9],[2]). For the same reason, attention is given to the action of the symmetric group; an action which, in particular, defines the subspaces of symmetric and anti-symmetric tensors.

The method we use was developed by Martens [5], and extended in [4]. The basic idea is to consider a filtration in $H^n = \underset{n}{\otimes} H$ according to the vanishing on the diagonal of $D \times \ldots \times D$. This gives the decomposition of T^n when T is strongly supported by the forward light cone ([4], Proposition 2.5). We paraphrase this result

*This paper is partially supported by NSF Grant number MCS 77-07596.

in §1. In §2 we show how to further extend this idea to cover the case of a wave-equation. A detailed knowledge of the corresponding representations is needed, and for this reason we restrict our attention to $G = SU(m,m)$. For the sake of simplicity we only treat the case of a scalar ("spin zero") wave-equation.

1. <u>Holomorphic discrete series</u>. To fix the notation and to introduce the method, we first consider the case in which the representation is strongly supported by the solid forward light cone (see [4] and below). The holomorphic discrete series are examples of such, but not all such are discrete series.

Let $G = SU(m,m)$ and let τ be an irreducible unitary representation of the maximal compact subgroup K of G in a finite dimensional vector space V_τ . The elements of K are of the form $\begin{pmatrix} a & -b \\ b & a \end{pmatrix}$; $(a+ib),(a-ib) \in U(m) \times U(m)$, and $\det(a+ib)(a-ib) = 1$. Let $u_1 = (a+ib)$ and $u_2 = (a-ib)$. To begin with we shall work with the unbounded realization $\mathcal{D}$ of G/K . Specifically $\mathcal{D} = \{z \in gl(m,\mathbb{C}) | \frac{z-z^*}{2i} \in C_+\}$, where C_+ is the space of strictly positive definite $m \times m$ matrices. Corresponding to τ there is a representation U_τ of G , holomorphically induced from τ , which acts on the space $\mathcal{O}(\mathcal{D},V_\tau)$ of holomorphic functions $f: \mathcal{D} \to V_\tau$ by

1.1 $$(U_\tau(g)f)(z) = J_\tau(g^{-1},z)^{-1}f(g^{-1}z) .$$

If $g = \begin{pmatrix} a & b \\ c & d \end{pmatrix} \in G$, $gz = \frac{az+b}{cz+d}$, and if $k \in K$, $J_\tau(k,i) = J_\tau(k) = \tau(k)$. U_τ is said to be supported by the solid forward light cone C^+ if it is unitary in a Hilbert space $H(\tau) \subset \mathcal{O}(\mathcal{D},V_\tau)$ and if the elements of $H(\tau)$ are Fourier-Laplace transforms of functions from C^+ to V_τ. If $H(\tau)$ is a reproducing kernel Hilbert space, it is easy to see that there exists a continuous function $F_\tau\colon C^+ \to \mathrm{Hom}(V_\tau)$ such that the kernel K_τ is given by

1.2 $$K_\tau(z,w) = \int_{C^+} F_\tau(k)e^{i\,\mathrm{tr}\,(z-w^*)k}dk .$$

If, for all $k \in C^+$, $F_\tau(k)$ is nonsingular, we say that U_τ is strongly supported by C^+.

Assume that U_τ is strongly supported by C^+. Consider

1.3 $$U_\tau^n = \bigotimes_n U_\tau = U_\tau \otimes \ldots \otimes U_\tau .$$

This representation acts on the space of holomorphic functions $f(z_1,\ldots,z_n)$ from $\mathcal{D} \times \ldots \times \mathcal{D}$ to $\bigotimes_n V_\tau$ in the obvious way. It is unitary on the subspace $H^n(\tau) = \bigotimes_n H(\tau)$. We consider its restriction to this space. Introduce new variables

1.4 $y_1 = z_1 + \ldots + z_n$, and $y_i = z_1 - z_i$ for $i = 2,\ldots,n$.

Recall that if $g = \begin{pmatrix} a & b \\ c & d \end{pmatrix} \in G$, then ([4])

$$1.5 \qquad (gz_1 - gz_2) = (z_2c^* + d^*)^{-1}(z_1 - z_2)(cz_1 + d)^{-1} \quad .$$

The representation U^n_τ is the restriction of a representation of $G \times \ldots \times G$ to the diagonal subgroup G_1 $(\simeq G)$. Its decomposition is given by [4], Proposition 2.5. We rephrase this result:

Let, for $j = 0,1,2,\ldots$,

$$1.6 \quad H^n_j(\tau) = \{f \in H^n(\tau) \mid (\partial/\partial y_2)^{\alpha_2} \ldots (\partial/\partial y_n)^{\alpha_n} f(z,\ldots,z) = 0$$

$$\text{for all } |\alpha| = \alpha_2 + \ldots + \alpha_n \leq j\} \; .$$

Here, for each y_i and α_i , $(\partial/\partial y_i)^{\alpha_i}$ runs through the set of all constant coefficient differential operators of degree α_i in the m^2 entries of y_i . The spaces $H^n_j(\tau)$ form a decreasing sequence of closed, invariant subspaces of $H^n(\tau)$. To be consistent we let $H^n(\tau) = H^n_{-1}(\tau)$. The functions in $H^n_j(\tau)$ are, to the lowest order in $y_2,\ldots,y_n$, homogeneous polynomials of degree $j+1$.

<u>Definition 1.1</u>. $\mathbb{C}^{n-1}_\tau[y_2,\ldots,y_n] = \mathbb{C}^{n-1}_\tau$ denotes the set of polynomial functions $p\colon (y_2,\ldots,y_n) \to \underset{n}{\otimes} V_\tau$. $\mathbb{C}^{n-1}_\tau(j)$ denotes the subspace of homogeneous polynomials of degree j .

K acts on $\mathbb{C}_\tau^{n-1}$ by $J_{\tau,n}$:

1.7 $(J_{\tau,n}(k)p)(y_2,\ldots,y_n) = (\tau^n(k)p)(u_2^{-1}y_2u_1,\ldots,u_2{}^{-1}y_nu_1)$.

Here $\tau^n = \underset{n}{\otimes}\tau$, and $k = (u_1,u_2)$. Let $\underset{i\in I}{\oplus}\rho_i(\tau,n)$ be the decomposition of $J_{\tau,n}$ into irreducible representations (there may be multiplicities) and let $U_{\rho_i(\tau,n)}$ denote the representation of G holomorphically induced from $\rho_i(\tau,n)$.

<u>Proposition 1.2</u> ([4]). If U_τ is strongly supported by C^+ ,

1.8 $$U_\tau^n = \underset{i\in I}{\oplus} U_{\rho_i(\tau,n)} .$$

Analogously to Definition 1.1 we can define $\mathbb{C}_\tau^n(j)$ for the variables $z_1,\ldots,z_n$. K is represented on this space by an action similar to (1.7).

<u>Definition 1.3</u>. T_j^n denotes the subspace of $\mathbb{C}_\tau^n(j)$ that is annihilated by $(\partial/\partial y_1) = (\partial/\partial z_1 + \ldots + \partial/\partial z_n)$.

T_j^n is clearly a submodule which, in a natural way, is isomorphic to $\mathbb{C}_\tau^{n-1}(j)$.

If $I[y_1]$ denotes the ideal in $\mathbb{C}_\tau^n$ generated by the entries of y_1 , T_j^n is a complement in $\mathbb{C}_\tau^n(j)$ of $\mathbb{C}_\tau^n(j) \cap I[y_1]$. Thus, if $\mathbb{C}_0^1[y_1](i) = \mathbb{C}_0^1(i)$ denotes the space of $\mathbb{C}$-valued homogeneous polynomials of degree i , we have

Lemma 1.4.

1.9 $$\mathbb{C}^n_\tau(j) = \bigoplus_{i=0}^{j} \mathbb{C}^1_0(i)\cdot T^n_{j-i} \quad .$$

K acts on $\mathbb{C}^1_0$ by, for $q \in \mathbb{C}^1_0$ and $k = (u_1,u_2)$,

1.10 $$(k\cdot q)(y_1) = q(u_2^{-1}y_1u_1) \quad .$$

It is clear that K leaves $\mathbb{C}^1_0(i)\cdot T^n_{j-i}$ invariant, and that its action on this space is given as the tensor product of (1.10) with its action on T^n_{j-i} . Equation (1.9) is then an equality between K-modules.

Let us return to the space $H^n(\tau)$. On this space the symmetric group $\mathfrak{S}(n)$ acts by

1.11 $$\sigma\,(f_1 \otimes \dots \otimes f_n) = f_{\sigma(1)} \otimes \dots \otimes f_{\sigma(n)} \quad .$$

This action is the tensor product of $\mathfrak{S}(n)$'s natural action on $\otimes_n V_\tau$ with its action on scalar valued functions on $\mathcal{B} \times \dots \times \mathcal{B}$. The spaces $H^n_j(\tau)$ are invariant, and thus $\mathfrak{S}(n)$ acts on $\mathbb{C}^{n-1}_\tau(j)$, for each j , and hence, by isomorphism, on T^n_j . $\mathfrak{S}(n)$ also acts on $\mathbb{C}^n_\tau$ by (1.11), and the above action on T^n_j is in fact, as follows from these remarks, the restriction of this latter action to that subspace. Observe that the direct sum decomposition (1.9) is invariant under the action of $\mathfrak{S}(n)$.

Let $\mathbb{C}_\tau^n(j)_s$ denote the subspace of $\mathbb{C}_\tau^n(j)$ that transforms under $\mathfrak{S}(n)$ according to a given symmetry s , and let $(T_j^n)_s$, $(T^n)_s$, and $H^n(\tau)_s$ be defined analogously, where $T^n = \bigoplus_{j=0}^{\infty} T_j^n$. Then, if $\bigoplus_{i\in I_s} \rho_i(\tau,n)$ denotes the decomposition of K's action on $(T^n)_s$ into irreducible representations, and if $(U_\tau^n)_s$ denotes the restriction of U_τ^n to $H^n(\tau)_s$,

Theorem 1.5. $(U_\tau^n)_s = \bigoplus_{i\in I_s} U_{\rho_i(\tau,n)}$. The K-types of $(T^n)_s$ can be obtained from the K-types of $(\mathbb{C}_\tau^n)_s$ by the recursion formula

$$1.12 \quad (T_j^n)_s = \mathbb{C}_\tau^n(j)_s - \mathbb{C}_0^1(1)\cdot(T_{j-1}^n)_s - \dots - \mathbb{C}_0^1(n)\cdot(T_0^n)_s \quad .$$

2. Wave-equations. Let us for simplicity consider the constant coefficient differential operator $\det(\mathbb{D}) = \det(\partial/\partial z)$ which is the (formal) Fourier-Laplace transform of the function $\det k$ on $\overline{C^+}$ (cf. [3], §2). There are other "Dirac-type" matrix valued constant coefficient differential operators which can be treated analogously. Take $m \geq 2$.

Let W denote the representation of G on the space of holomorphic functions on $\mathcal{D}$ given by, for

$g^{-1} = \begin{pmatrix} a & b \\ c & d \end{pmatrix} \in G$,

$$\text{2.1} \qquad (W(g)f)(z) = \det(cz+d)^{1-m} f(g^{-1}z) \; .$$

This representation corresponds to a discrete point in the Wallach set ([10]) beyond the analytic continuation. It is known from [7] (see also [3]) that there exists a reproducing kernel Hilbert space H of holomorphic solutions to $\det(\mathbb{D})\varphi = 0$ on $\mathcal{D}$, in which W is unitary. In fact ([7], §4.5-4.6), there exists a semi-invariant measure μ on the set $b(C^+) = \{k \in gl(m,\mathbb{C}) \mid k \geq 0 \text{ and } \det k = 0\}$ such that $f \in H$ if and only if

$$\text{2.2} \qquad f(z) = \int_{b(C^+)} e^{i\,tr(zk)} \varphi(k) d\mu(k)$$

for some $\varphi \in L^2(b(C^+), d\mu)$. As before we let $\mathbb{C}_0^1 = \mathbb{C}_0^1[z]$ denote the space of polynomials in the m^2 entries of z .

Definition 2.1. $I(\det z)$ denotes the ideal in $\mathbb{C}_0^1$ generated by $\det z$.

Remark: $I(\det z)$ is clearly a prime ideal.

Let $\mathbb{E}_0^1 = \mathbb{C}_0^1[\partial/\partial z]$ denote the space of constant coefficient differential operators in the entries of z , and let $I(\det \mathbb{D})$ be defined analogously. There is a natural pairing between $\mathbb{E}_0^1$ and $\mathbb{C}_0^1$ defined as

$$2.3 \qquad \langle E_{j_1}, p_{j_2}\rangle = E_{j_1}(p_{j_2})(o)$$

if $E_{j_1} \in \mathbb{E}_0^1(j_1)$ and $p_{j_2} \in \mathbb{C}_0^1(j_2)$. When $j_1 = j_2$, (2.3) of course gives the canonical duality between $\mathbb{E}_0^1(j_1)$ and $\mathbb{C}_0^1(j_1)$.

Lemma 2.2. Let $p \in \mathbb{C}_0^1$. Then $(\det \mathbb{D})p = 0$ if and only if $\langle I(\det \mathbb{D}), p\rangle = 0$.

Proof: If $\langle I(\det \mathbb{D}), p\rangle = 0$, all derivatives of $(\det \mathbb{D})p$ vanish at 0, and hence $(\det \mathbb{D})p = 0$. The converse is obvious.

We can expand any function in H in a power series at the point $i \in \mathcal{D}$. The degree of the leading term is then K-invariant. Suppose $p_j(z-i)f(z) \in H$, that $f(i) \neq 0$, and that p_j is a homogeneous polynomial of degree j. Since $\det \mathbb{D}(p_j(z-i)f(z)) = 0$, it follows, by comparing degrees, that $\det \mathbb{D}(p_j(z-i)) = 0$, and hence $(\det \mathbb{D})p_j = 0$. Under $W(k)$'s action on H, for $k = (u_1,u_2) \in K$, $p_j(z-i)$ is mapped into $\det u_1^{m-1} p_j(u_2^{-1}(z-i)u_1)$.

Lemma 2.3. If $p_j \in \mathbb{C}_0^1(j)$ and if $(\det \mathbb{D})p_j = 0$, there exists a holomorphic function f such that $f(i) \neq 0$, and $p_j(z-i)f(z) \in H$.

<u>Proof</u>: Consider a function F_φ in H of the form

2.4 $$F_\varphi(z) = \int_{b(C^+)} e^{i\,\mathrm{tr}\,(zk)}\varphi(k)d\mu(k) \quad .$$

If F_φ is to be of the form $F_\varphi(z) = p_j(z-i)f(z)$, $f(i) \neq 0$, clearly

2.5 (i) $$\int_{b(C^+)} e^{-\,\mathrm{tr}\,k}\, q_\alpha(k)\varphi(k)d\mu(k) = 0$$

for all polynomials q_α of degree $\alpha < j$,

(ii) $$\int_{b(C^+)} e^{-\,\mathrm{tr}\,k}\, q_j(k)\varphi(k)d\mu(k) \neq 0$$

for some polynomial q_j of degree j .

It is easy to see that any function of the form $\varphi(k) = e^{-\,\mathrm{tr}\,k}q(k)$, where $q(k)$ is a polynomial, is in $L^2(b(C^+),d\mu)$. Moreover, $\|e^{-\,\mathrm{tr}\,k}q(k)\|_2 = 0$ if and only if q is zero on $b(C^+)$. A straightforward orbit argument now implies that q is zero on the set of points where $\det k = 0$. But this set is the locus for the ideal $I(\det k)$, hence

2.6 $$\|e^{-\,\mathrm{tr}\,k}q(k)\|_2 = 0 \Longleftrightarrow q \in I(\det k) \;.$$

Assume that $\varphi(k) = e^{-\,\mathrm{tr}\,k}q(k)$, $q(k) \notin I(\det k)$. Then

$$2.7 \qquad F_\varphi(z) = \int_{b(C^+)} e^{i\,\mathrm{tr}\,(z+i)k} q(k)\,d\mu(k)$$

$$= E \int_{b(C^+)} e^{i\,\mathrm{tr}\,(z+i)k} d\mu(k)$$

$$= E \det (z+i)^{1-m} ,$$

where E, except for a transposition, is $q(\partial/\partial z)$. In particular, $E \notin I(\det \mathbb{D})$. Conversely, if $E \notin I(\det \mathbb{D})$, then $q(k) \notin I(\det k)$. The proof is completed by an easy Gram-Schmidt argument.

It is now convenient to introduce the bounded version $\mathcal{B}$ of G/K; $\mathcal{B} = \{z \in gl(m,\mathbb{C}) | \; zz^* < 1\}$. The differential operator $\det \mathbb{D}$ can, in an obvious way, be extended from $\mathcal{B}$ to $gl(m,\mathbb{C})$, and then restricted to act on holomorphic functions on $\mathcal{B}$. The map C:

$$2.8 \qquad (Cf)(z) = (\det(i(z+1)))^{1-m} f\left(\frac{z-1}{i(z+1)}\right)$$

maps H unitarily onto a reproducing kernel Hilbert space H_b of holomorphic functions on $\mathcal{B}$, and it is easy to see that H_b consists of solutions to $(\det \mathbb{D})\varphi = 0$. We let $W_b = CWC^{-1}$ be the corresponding representation of G and observe that if $k = (u_1, u_2) \in K$,

$$2.9 \qquad (W_b(k)f)(z) = \det u_1{}^{m-1} f(u_2^{-1} z u_1) .$$

Corollary 2.4. f is a K-finite vector in H_b if and only if it is a polynomial and $(\det \mathbb{D})f = 0$.

Let λ' denote the representation of K in $\mathbb{E}_0^1$ which is the contragredient of the representation (2.9) of K in $\mathbb{C}_0^1$. If $k = (u_1,u_2) \in K$ define $\tilde{k} = (u_2,u_1) \in K$. λ' is then equivalent to the representation $k \to W_b(\tilde{k})$. This has the following straightforward consequence:

Corollary 2.5. The space $Q^1(\det \mathbb{D})$ of K-finite solutions to $(\det \mathbb{D})\varphi = 0$ is a complement to $I(\det z)$ in $\mathbb{C}_0^1$.

Since, for all i and j in $\mathbb{N}$, $\det z^i \cdot \mathbb{C}_0^1(j) \subset \mathbb{C}_0^1$, we may state

Corollary 2.6.

2.10 $$Q^1(\det \mathbb{D}) = \mathbb{C}_0^1 - \bigoplus_{j=0}^{\infty} \bigoplus_{i=1}^{\infty} (\det z)^i \cdot \mathbb{C}_0^1(j) \quad .$$

For $n \geq 2$ we let $W^n = \otimes_n W_b$ and $H^n = \otimes_n H_b$. Even though the domain has changed, we can keep the notation from §1. A function in H_j^n is then of the form

2.11 $$p_{j+1}(y_2,\ldots,y_n)f(z_1,\ldots,z_n) ,$$

where p_{j+1} is a homogeneous polynomial of degree $j+1$.

Let us assume that f is not identically zero on the diagonal $z_1 = \ldots = z_n$. It follows that

2.12 $$(\det \mathbb{D}_{z_i})p_{j+1} = 0 \quad \text{for } i = 1,\ldots,n .$$

Let $Q^n(\det \mathbb{D})$ denote the subset of $\mathbb{C}_0^n$,

2.13 $$Q^n(\det \mathbb{D}) = \{p \in \mathbb{C}_0^n \mid (\det \mathbb{D}_{z_i})p = 0 \text{ for } i = 1,\ldots,n\} ,$$

and let

2.14 $$P^n = Q^n(\det \mathbb{D}) \cap T^n .$$

P^n is invariant under K (2.9). Let $\bigoplus_{j\in J} \mu_j^n$ denote the decomposition of K's action on P^n into irreducible representations.

<u>Theorem 2.7</u>. $W^n = \bigoplus_{j\in J} U_{\mu_j^n}$.

<u>Proof</u>: It follows from (2.12) and §1 that W^n is contained in the right hand side. To prove the equality, we need only observe that P^n is contained in H^n (c.f. the proof of Proposition 2.5 in [4]), and this follows from Corollary 2.4.

Let $I^n(\det z)$ denote the ideal in $\mathbb{C}_0^n$ generated by the functions $\det z_1, \ldots, \det z_n$ and let, as before, $I[y_1]$ denote the ideal generated by the entries of $y_1 = z_1 + \ldots + z_n$. As a consequence of Corollary 2.5 and the above we have

Proposition 2.8. The decomposition into irreducible representations of the action of K on the complement to $I^n(\det z) + I[y_1]$ in $\mathbb{C}_0^n$ is given by $\bigoplus_{j \in J} \mu_j^n$ (as above).

The following is elementary (cf. [1], p. 22)

Lemma 2.9. $I^n(\det z)$ is prime.

Corollary 2.10. Let $q \in \mathbb{C}_0^n$, $n \geq 2$. If $q \cdot \det(z_1 + \ldots + z_n)$ is in $I^n(\det z)$, then q is in $I^n(\det z)$.

Proof: Since $I^n(\det z)$ is prime, either q or $\det(z_1 + \ldots + z_n)$ is in the ideal. If $n \geq 2$, clearly $\det(z_1 + \ldots + z_n)$ is not $(m \geq 2)$.

Lemma 2.11. For $n \geq 3$,

$$I^n(\det z) \cap I[y_1] = I^n(\det z) \cdot I[y_1]. \quad (2.15)$$

Proof: Let $f_1, \ldots, f_n \in \mathbb{C}_0^n$ and assume that $f_1 \det z_1 + \ldots + f_n \det z_n \in I[y_1]$. Introduce variables

$z_1,\ldots,z_{n-1},y_1$. Each f_i $(i = 1,\ldots,n)$ can be written as $f_i = \alpha_i + \beta_i$, where $\beta_i \in I[y_1]$ and α_i is a polynomial in $z_1,\ldots,z_{n-1}$. It follows that

2.16 $$f_1 \det z_1 + \ldots + f_{n-1} \det z_{n-1} + f_n \det z_n =$$

$$\alpha_1 \det z_1 + \ldots + \alpha_{n-1} \det z_{n-1} + \alpha_n \det (z_1+\ldots+z_{n-1}) + \alpha_n q_1 + q_2 ,$$

where $q_1 \in I[y_1]$ and $q_2 \in I^n(\det z)\cdot I[y_1]$. Thus,

$$\alpha_1 \det z_1 + \ldots + \alpha_{n-1} \det z_{n-1} + \alpha_n \det (z_1 + \ldots + z_{n-1})$$

belongs to $I[y_1]$, and hence is zero. By Corollary 2.10 α_n is in the ideal generated by $\det z_1,\ldots,\det z_{n-1}$, and hence $\alpha_n \cdot q_1 \in I^n(\det z)\cdot I[y_1]$.

$Q^n = Q^n(\det \mathbb{D})$ is a complement to $I^n(\det z)$ in $\mathbb{C}_0^n$ by Corollary 2.5. We let Q_j^n denote the subspace of Q^n of homogeneous polynomials of degree j , and let

2.17 $$P_j^n = Q^n \cap T_j^n .$$

From Lemma 2.11 we conclude

<u>Proposition 2.12</u>. For $n \geq 3$,

2.18 $$Q_j^n = \bigoplus_{i=0}^{j} \mathbb{C}_0^1(i)\cdot P_{j-1}^n .$$

This formula is based on the decomposition

2.19 $$\mathbb{C}_0^n - I^n(\det z) - I[y_1] + I^n(\det z)\cdot I[y_1] =$$

$$\mathbb{C}_0^n - I^n(\det z) - I[y_1](\mathbb{C}_0^n - I^n(\det z)) .$$

An equivalent formula can be obtained by using the decomposition

2.20 $$\mathbb{C}_0^n - I[y_1] - I^n(\det z)(\mathbb{C}_0^n - I[y_1]) .$$

The decomposition (2.18) of Q_j^n is, just as (1.9), invariant under the action of $\mathbf{S}(n)$. With it, P_j^n can be expressed in terms of tensor products of the spaces Q_j^n and $\mathbb{C}_0^1(i)$. Moreover, for each j , Q_j^n can be obtained recursively from the spaces $\mathbb{C}_0^1(i)$, in a manner analogous to the derivations of (1.9) and (2.10).

As for the case $n = 2$, observe that

2.21 $$I^2(\det z) + I[y_1] \doteq I(\det(z_1 - z_2)) + I[y_1]$$

and that

2.22 $$I(\det(z_1-z_2)) \cap I[y_1] = I(\det(z_1-z_2))\cdot I[y_1] .$$

Since $z_1 - z_2$ is equal to the variable y_2 we conclude:

Proposition 2.13. The space P^2 is, as a K-module, equivalent to the K-module Q^1 (Corollaries 2.5 and 2.6).

Let Q^1_e and Q^1_o denote the set of homogeneous polynomials of even and odd degrees, respectively, in Q^1, and let $\bigoplus_{i\in I_+} \gamma_i$ and $\bigoplus_{i\in I_-} \gamma_i$ denote the decomposition of K on Q^1_e and Q^1_o, respectively. Denote by H^2_+ and H^2_- the symmetric and antisymmetric subspaces of H^2_1 and define $W^2_\pm$ analogously.

Proposition 2.14. $W^2_\pm = \bigoplus_{i\in I_\pm} U_{\gamma_i}$.

3. Concluding remarks. The problem of decomposing tensor products of infinite-dimensional representations of $G = SU(m,m)$ has been reduced to the problem of decomposing tensor products of finite-dimensional representations of compact groups, e.g. decomposing $\mathbb{C}^n_\tau(j)$ under K. In the case where τ is a character, the decomposition of $\mathbb{C}^1_\tau$ was found by Schmid [8], for more general groups. A related study is that of Procesi [6]. In principle, then, the decomposition of $\mathbb{C}^n_\tau$, and more generally of $(\mathbb{C}^n_\tau)_s$, can be found from classical invariant theory. In practice, of course, this is quite hard, and we shall not get into this problem here.

We wish to express our gratitude to I. E. Segal and T. Washburn for friendly help and conversations.

REFERENCES

[1] R. Hartshorne, Algebraic Geometry, Springer-Verlag, New York, Heidelberg, Berlin, 1977.

[2] H. P. Jakobsen, B. Ørsted, I. E. Segal, B. Speh and M. Vergne, Symmetry and causality properties of physical fields, Proc. Natl. Acad. Sci. USA 75 (1978), 1609-1611.

[3] H. P. Jakobsen, Intertwining differential operators for $Mp(n,\mathbb{R})$ and $SU(n,n)$, to appear in Trans. Amer. Math. Soc.

[4] H. P. Jakobsen and M. Vergne, Restrictions and expansions of holomorphic representations, to appear in J. Functional Analysis.

[5] S. Martens, The characters of holomorphic discrete series, Proc. Natl. Acad. Sci. USA 72 (1975), 3275-3276.

[6] C. Procesi, The invariant theory of $n \times n$ matrices, Advances in Math. 19 (1976), 306-381.

[7] H. Rossi and M. Vergne, Analytic continuation of the holomorphic discrete series of a semi-simple Lie group, Acta Math. 136 (1976), 1-59.

[8] W. Schmid, Die randwerte holomorpher funktionen auf hermitesch symmetrischen räumen, Invent. Math. 9 (1969), 61-80.

[9] I. E. Segal, Mathematical Cosmology and Extragalactic Astronomy, Academic Press, New York, 1976.

[10] N. Wallach, Analytic continuation for the holomorphic discrete series, I and II, to appear in Trans. Amer. Math. Soc.

Department of Mathematics
Brandeis University
Waltham, Massachusetts 02154/USA

W-MODULE STRUCTURE IN THE PRIMITIVE SPECTRUM OF THE ENVELOPING ALGEBRA OF A SEMISIMPLE LIE ALGEBRA

A. JOSEPH

Department of Mathematics
Tel-Aviv University
Ramat Aviv
ISRAEL

This paper was written while the author was a guest of the Institute for Advanced Studies, The Hebrew University of Jerusalem and on leave of absence from the Centre National de la Recherche Scientifique, France. (August 1978).

ABSTRACT :

Formulae developed to give a positive answer to Dixmier's problem for Verma and principal series submodules are used to show that each primitive ideal in the enveloping algebra of a semisimple Lie algebra identifies with a left ideal in the group algebra of the Weyl group. The possible behaviour of these left ideals under right multiplication leads to a conjecture for the set of order relations in the primitive spectrum.

1 Introduction.

1.1 For each Lie algebra $\underline{a}$, let $U(\underline{a})$ denote the enveloping algebra of $\underline{a}$ and $Z(\underline{a})$ the centre of $U(\underline{a})$. Let Prim $U(\underline{a})$ denote the set of primitive ideals of $U(\underline{a})$ considered as an ordered set by inclusion of elements. Given M a $U(\underline{a})$ module of finite length, let $\underline{JH}(M)$ denote the set of simple factors of M (with multiplicities) and let $[M:L]$ denote the number of times L occurs in $\underline{JH}(M)$. Given V a left (resp. right) $U(\underline{a})$ module we write $\ell(V) := \{a \in U(\underline{a}) : aV = 0\}$, $r(V) := \{a \in U(\underline{a}) : Va = 0\}$.

1.2 Let $\underline{g}$ be a complex semisimple Lie algebra, $\underline{h}$ a Cartan subalgebra for $\underline{g}$ and W the Weyl group for the pair $\underline{g},\underline{h}$. The aim of this paper is to describe Prim $U(\underline{g})$ as an ordered set. This problem easily reduces to the description of each fibre $\underline{X}_{\hat{\lambda}} : \hat{\lambda} \in \text{Max } Z(g)$, for the map $I \to I \cap Z(\underline{g})$ of Prim $U(\underline{g})$ onto Max $Z(\underline{g})$, as an ordered set. Moreover after Duflo ([4], Thm.1) each fibre $\underline{X}_{\hat{\lambda}}$ is some homomorphic image of W and in particular has finite cardinality. Let $\underline{Y}_{\hat{\lambda}}$ denote the set of semiprime ideals obtained by taking all possible intersections of elements of $\underline{X}_{\hat{\lambda}}$.

Now consider the matrix of non-negative integers $[M:L]$ where M runs over the set of Verma modules with infinitesimal character $\hat{\lambda}$ and L the corresponding set of simple quotients. We call this the multiplicity matrix and use it to define a basis in $\mathbb{Q}W$. Our first and main result (Theorem 5.3) is that each $I \in \underline{Y}_{\hat{\lambda}}$ defines a left ideal of $\mathbb{Q}W$ spanned by the given basis. This implies a number of order relations in $\underline{X}_{\hat{\lambda}}$ including those given by Duflo ([4], III, Cor. 1) and those given in ([7], 5.1). Conversely we suggest

(5.7, Conjecture A) that each such left ideal corresponds to an element of $\underline{Y}_{\tilde{\lambda}}$. This would imply that the multiplicity matrix completely determines the set of order relations in $\underline{X}_{\tilde{\lambda}}$. Its proof should involve a method of separating ideals. From the work of Borho and Jantzen [1] , Vogan ([12] , Sect. 3) has developed a set of conditions (which we refer to as the BJV conditions) on the order relations in $\underline{X}_{\tilde{\lambda}}$ which are given in terms of the multiplicity matrix. In our language these correspond to the action of right multiplication in $\mathbb{Q}W$. Our second result (Theorem 5.9) shows that BJV conditions are satisfied by our conjectured solution. Consequently our conjecture is proven if one can show that the BJV conditions completely determine the order relations in $\underline{X}_{\tilde{\lambda}}$. Vogan has already shown that this is at least very nearly true.

1.3 The above asserted W-module structure of the Prim $U(\underline{g})$ fibres not only shows how the order relations in Prim $U(\underline{g})$ might be compactly expressed; but also "explains" some otherwise rather mysterious empirical facts about Prim $U(\underline{g})$ (such as the truth of the Jantzen conjecture ([9] , 5.3; [12] , Sect. 6)) . Its proof combines the method of coherent continuation with some of the formulae developed in [10] .

2 Preliminaries.

2.1 Unless otherwise specified all vector spaces are assumed over the complex field $\mathbb{C}$. For each vector space let $S(V)$ denote the symmetric algebra over V and V^* the dual of V .

2.2 With $\underline{g}$, $\underline{h}$, W as in 1.2 , let $R \subset h^*$ be the set of non-zero roots, $R^+ \subset R$ a system of positive roots, $B \subset R^+$ the corresponding set of simple roots, s_α the reflection corresponding to the root α, $P(R)$ the lattice of integral weights, ρ the half sum of the positive roots. Fix a Chevalley basis for $\underline{g}$ and let X_α denote the element of this basis of weight $\alpha \in R$. Set

$$\underline{n}^+ = \bigoplus_{\alpha \in R^+} \mathbb{C} X_\alpha , \quad \underline{n}^- = \bigoplus_{\alpha \in R^+} \mathbb{C} X_{-\alpha} .$$

Let P denote the projection of $U(\underline{g})$ onto $U(\underline{h})$ defined by the decomposition $U(\underline{g}) = U(\underline{h}) \oplus (\underline{n}^- U(\underline{g}) + U(\underline{g})\underline{n}^+)$.

2.3 Call $\lambda \in h^*$ *dominant* if $2(\lambda,\alpha)/(\alpha,\alpha) \notin \mathbb{N}^-$ for all $\alpha \in R^+$ and *regular* if $(\lambda,\alpha) \neq 0$, for all $\alpha \in R$.
For each $\lambda \in \underline{h}^*$, set $R_\lambda = \{\alpha \in R : 2(\lambda,\alpha)/(\alpha,\alpha) \in \mathbb{Z}\}$, $R_\lambda^+ = R^+ \cap R_\lambda$, $B_\lambda \subset R_\lambda^+$ the corresponding set of simple roots, W_λ the subgroup of W generated by the $s_\alpha : \alpha \in R_\lambda$. For each $B' \subset B_\lambda$, let $W_{B'}$ be the subgroup of W generated by the s_α : $\alpha \in B'$ and let $w_{B'}$ be the unique element of $W_{B'}$ taking B' to $-B'$. When $B' = B_\lambda$, we write $w_{B'} = w_\lambda$.

2.4 For each $\lambda \in \underline{h}^*$, let $M(\lambda)$ denote the Verma module (with highest weight $\lambda - \rho$) associated with the quadruplet $\underline{g}, \underline{h}, B, \lambda$ ([2], 7.1.4). Let $L(\lambda)$ denote the unique simple quotient of $M(\lambda)$ and set $J(\lambda) = \text{Ann } M(\lambda)$. Let $\hat{\lambda}$ denote the orbit of λ under W which by ([2], 7.4.7) identifies with an element of $\text{Max } Z(\underline{g})$. Then by ([4], Thm. 1) we may write

$\underline{X}_{\hat{\lambda}} = \{J(\mu) : \mu \in \hat{\lambda}\}$.

2.5 Let $u \to {}^t u$ (resp. $u \to \check{u}$) denote the antiautomorphism of $U(\underline{g})$ defined through ${}^t X_\alpha = X_{-\alpha} : \alpha \in R$; ${}^t H = H : H \in \underline{h}$ (resp. $\check{X} = -X : X \in \underline{g}$). Identify $U := U(\underline{g}) \otimes U(\underline{g})$ canonically with $U(\underline{g} \times \underline{g})$ and set $\underline{k} = \{(X, -{}^t X) : X \in g\}$.

Let M, N be left $U(\underline{g})$ modules and define $\mathrm{Hom}_{\mathbb{C}}(M, N)$ as a U module through $((a \otimes b).x)m = {}^{t\vee}a x \check{b} m$: $a, b \in U(\underline{g})$, $x \in \mathrm{Hom}_{\mathbb{C}}(M,N)$, $m \in M$. Let $L(M,N)$ denote the subspace of $\mathrm{Hom}_{\mathbb{C}}(M,N)$ of all $\underline{k}$-finite elements (which is again a U module). Consider $(M(-\lambda) \otimes M(-\mu))^*$ as a U module by transposition and let $L(\lambda,\mu)$ denote the subspace of all $\underline{k}$-finite elements (which is again a U module). Let $V(\lambda,\mu)$ denote the unique simple subquotient of $L(\lambda,\mu)$ in which the simple $\underline{k}$ module with extreme weight $\lambda - \mu$ occurs exactly once.

2.6 Identify $U(\underline{h})$ with $S(\underline{h})$ and hence with the set of polynomial functions on $\underline{h}^*$. For each $\lambda \in \underline{h}^*$, define $P_\lambda : U(\underline{g}) \to \mathbb{C}$ through $P_\lambda(a) = (P(a), \lambda - \rho)$. As noted by Duflo ([4], Sect. I) one has

$$J(\lambda) = \{a \in U(\underline{g}) : P_\lambda({}^t vau) = 0 \text{ , for all } u, v \in U(\underline{g})\} .$$

For each ideal K of $S(\underline{h})$, let $\underline{V}(K) \subset h^*$ denote its zero variety. Following ([7], Sect. 2) we define the *characteristic variety* of an ideal I of $U(\underline{g})$ through $\underline{V}(I) := \underline{V}(P(I)) + \rho$, and for each $\lambda \in \underline{h}^*$ we write $\underline{V}(J(\lambda)) = \underline{V}(\lambda)$.

LEMMA - ([7], 2.1). *For each* $\lambda, \mu \in \underline{h}^*$, *one has* $\mu \in \underline{V}(\lambda)$ *if and only if* $J(\lambda) \subset J(\mu)$.

2.7 By 2.6 a knowledge of the $\underline{V}(\mu) : \mu \in \hat{\lambda}$ is equivalent to a knowledge of $\underline{X}_{\hat{\lambda}}$ as an ordered set. Actually by virtue of ([7], 4.2) it suffices to fix $-\lambda \in \underline{h}^*$ dominant and to determine the *reduced characteristic variety* $\underline{V}_\lambda(w\lambda) := \underline{V}(w\lambda) \cap W_\lambda \lambda$ of each $J(w\lambda) : w \in W_\lambda$. More generally let $\underline{Z}_{\hat{\lambda}}$ denote the set of ideals of $U(\underline{g})$ whose intersection with $Z(\underline{g})$ is just the maximal ideal defined by $\hat{\lambda}$. Then $\underline{X}_{\hat{\lambda}}$ (resp. $\underline{Y}_{\hat{\lambda}}$) is the subset of $\underline{Z}_{\hat{\lambda}}$ of all prime (resp. semiprime) ideals. Define a map $\underline{V}_{\hat{\lambda}} : \underline{Z}_{\hat{\lambda}} \rightarrow \underline{P}(W_\lambda \lambda)$ through $\underline{V}_{\hat{\lambda}}(I) = \underline{V}(I) \cap W_\lambda \lambda$. Then (c.f. [7], 2.1)

LEMMA - *For all* $I, J \in \underline{Z}_{\hat{\lambda}}$,

(i) $\underline{V}_\lambda(\sqrt{I}) = \underline{V}_\lambda(I)$.

(ii) $\underline{V}_\lambda(I) \subset \underline{V}_\lambda(J)$ *if and only if* $I \supset J$ *and the restriction of* $\underline{V}_\lambda$ to $\underline{Y}_{\hat{\lambda}}$ *is injective.*

(iii) $\underline{V}_\lambda(I \cap J) = \underline{V}_\lambda(I) \cup \underline{V}_\lambda(J)$.

(iv) $\underline{V}_\lambda(I + J) = \underline{V}_\lambda(I) \cap \underline{V}_\lambda(J)$.

It follows that the knowledge of $\underline{X}_{\hat{\lambda}}$ as an ordered set is equivalent to a knowledge of $\operatorname{Im} \underline{V}_\lambda$. In fact $\underline{X}_{\hat{\lambda}}$ is exactly the intersection of $\underline{Y}_{\hat{\lambda}}$ with the preimage of the set of indecomposable elements of $\operatorname{Im} \underline{V}_\lambda$ (i.e. those elements of $\operatorname{Im} \underline{V}_\lambda$ which are not the union of distinct elements of $\operatorname{Im} \underline{V}_\lambda$).

2.8 Set $\underline{b} = \underline{h} \oplus \underline{n}^+$. For each $\lambda \in \underline{h}^*$, let $\underline{M}$ (resp. $\underline{M}_{\hat{\lambda}}$) denote the category of left $U(\underline{g})$ modules generated by a finite dimensional $\underline{b}$ module which is $\underline{h}$ diagonalizable (resp. and with infinitesimal character $\hat{\lambda}$) . One has $M(w\lambda)$, $L(w\lambda) \in \underline{M}_{\hat{\lambda}}$, for all $w \in W$. Each $M \in \underline{M}$ admits a formal character ch M defined as in ([2], 7.5.2) .

2.9 A U module is said to be *admissable* if it can be written as a direct sum of finite dimensional $\underline{k}$ primary components. Let $\underline{L}$ denote the category of admissable U modules. Given $M , N \in \underline{M}$ one has $L(M,N) \in \underline{L}$ (c.f. [8], 4.3) .

2.10 Fix $-\lambda \in \underline{h}^*$ dominant and regular. For each $\alpha \in B_\lambda$, choose $\mu_\alpha \in P(R)$ such that $\mu_\alpha - \lambda$ is dominant and $(\beta , \mu_\alpha - \lambda) = 0$: $\beta \in R^+$ implies $\beta = \alpha$ (c.f. [1], 2.14) . Let E_α denote the finite dimensional simple $U(\underline{g})$ module with extreme weight μ_α and E_α^* its contragredient module.

If M is a left $U(\underline{g})$ module we define following Vogan ([12], Sect.2)

$$Q_\lambda(M) = \{m \in M : \text{for all } x \in \hat{\lambda} \text{, there exists } r \in \mathbb{N} \text{ with } x^r m = 0\} ,$$

$$\varphi_\alpha(M) = Q_\lambda(Q_{\lambda-\mu_\alpha}(M) \otimes E_\alpha) ,$$
$$\Psi_\alpha(M) = Q_{\lambda-\mu_\alpha}(Q_\lambda(M) \otimes E_\alpha^*) .$$

Given $M \in \underline{M}$, then $\varphi_\alpha(M) \in \underline{M}$. Again each U module L is in particular a left $U(\underline{g})$ module and so $\varphi_\alpha(L)$ is defined. By purely formal calculations (c.f. [8], 7.2) we obtain

LEMMA - *For each* $M,N \in \underline{M}$,

(i) $\varphi_\alpha(L(M,N))$ *identifies with a submodule of* $L(M,\varphi_\alpha N)$.

(ii) $r(\varphi_\alpha(L(M,N))) \supset r(L(M,N))$.

Similar assertions hold for φ_α replaced by Ψ_α .

3 The Multiplicity Matrix.

From now on we fix $-\lambda \in \underline{h}^*$ dominant and regular and set $\overline{w} = w_\lambda w^{-1} w_\lambda$ for all $w \in W_\lambda$. (We remark that by ([1], 2.12) it suffices to determine $\underline{X}_{\overline{\lambda}}$ as an ordered set for λ regular.

3.1 Let $\leqslant$ denote the Bruhat ordering on W_λ as defined in ([2], 7.7.3). The *multiplicity matrix* b_λ is defined to have entries $b_\lambda(w,w') : w,w' \in W_\lambda$ given by

$$b_\lambda(w,w') := [M(w\lambda) : L(w'\lambda)] .$$

Actually b_λ depends only weakly on λ. Indeed given $\nu \in P(R)$ such that $-(\lambda+\nu)$ is dominant and regular one has ([6], 2.15) that $b_\lambda = b_{\lambda+\nu}$ and for this reason we shall generally drop the subscript. Again for all $w,w' \in W_\lambda$, one has $b(w,w) = 1$ and $b(w,w') \neq 0$, If and only if $w \geqslant w'$ (c.f.[2], 7.7.7). Consequently b can be considered as an upper triangular matrix with ones on the diagonal and we denote its inverse matrix by a. One has $a(w,w) = 1$ and $a(w,w') \neq 0$ if and only if $w \geqslant w'$.

3.2 The following result is a refinement of ([10], 5.4).

THEOREM - *Fix* $-\lambda, -\mu \in \underline{h}^*$ *dominant and regular with* $\lambda-\mu \in P(R)$. *Then for all* $w,w' \in W_\lambda$,

$$[L(-w\mu, -w_\lambda\lambda) : V(-w'\mu, -w_\lambda\lambda)] = [M(w\mu) : L(w'\mu)] .$$

Indeed

$$\underline{JH}(L(M(w_\lambda\lambda), M(w\mu)) = \underline{JH}(L(M(w^{-1}w_\lambda\lambda), M(\mu))), \quad \text{by ([10], 4.10(i))},$$

$= \underline{JH}(L(-\mu,-w^{-1}w_\lambda\lambda))$, by ([10], 3.2),

$= \underline{JH}(L(-w\mu,-w_\lambda\lambda))$, by ([3], III, 5.5).

Hence the assertion of the theorem follows from ([10], 3.5, 4.7).

3.3 COROLLARY - *Fix* $-\lambda \in h^*$ *dominant and regular. Then for all* $w, w' \in W_\lambda$,

(i) $a(\overline{w},\overline{w}') = a(w,w')$.

(ii) $b(\overline{w},\overline{w}') = b(w,w')$.

It is enough to establish (ii). Recall ([3], I, 4.1; III, 5.5) that for all $w \in W_\lambda$, $V(-w\mu,-\lambda) = V(-\mu,-w^{-1}\lambda)$, up to isomorphism and $\underline{JH}(L(-w\mu,-\lambda)) = \underline{JH}(L(-\mu,-w^{-1}\lambda))$. Then

$b(w,w') = [L(-w\mu,-w_\lambda\lambda) : V(-w'\mu,-w_\lambda\lambda)]$, by 3.2,

$= [L(-\mu,-w^{-1}w_\lambda\lambda) : V(-\mu,-w'^{-1}w_\lambda\lambda)]$, as above,

$= [L(-w^{-1}w_\lambda\lambda,-\mu) : V(-w'^{-1}w_\lambda\lambda,-\mu)]$, by transposition,

$= b(\overline{w},\overline{w}')$, by 3.2.

<u>Remarks</u>. Note this also gives a new proof of the asserted independence of $b(w,w')$ on λ. Had we taken λ dominant (and regular) then we would have obtained $b(w,w') = b(w^{-1}, w'^{-1})$, which is neater; but this convention is inconvenient in Section 4. Again the above result interrelates ([5], Thm. 2 (ii)) and ([5], Satz 11) and extends the validity of the latter.

4 The Multiplicity Basis for $\mathbb{Q}W_\lambda$.

4.1 For each $w \in W_\lambda$, set $S_\lambda(w) = \{\alpha \in R^+_\lambda : w\alpha \in R^-_\lambda\}$, $\tau_\lambda(w) = S_\lambda(w) \cap B_\lambda$. The *Duflo ordering* $\subseteq$ on W_λ is defined through $w \subseteq w'$ given $S_\lambda(w) \subseteq S_\lambda(w')$.

4.2 (Notation 3.1) . For each $w \in W_\lambda$, set

$$a(w) = \sum_{w' \in W_\lambda} a(w,w')\, w' .$$

Clearly $\{a(w) : w \in W_\lambda\}$ is a basis for $\mathbb{Q}W_\lambda$. It will play a central role in our analysis. First define for all $w,w' \in W_\lambda$, $\alpha \in B_\lambda$ the integers $c_\alpha(w,w')$, $d_\alpha(w,w')$ through

$$a(w)s_\alpha = \sum_{w' \in W_\lambda} c_\alpha(w,w')\, a(w') ,$$

$$s_\alpha a(w) = \sum_{w' \in W_\lambda} d_\alpha(w,w')\, a(w') .$$

LEMMA - *For all* $w,w' \in W_\lambda$, $\alpha \in B_\lambda$ *one has* $d_\alpha(\overline{w},\overline{w}') = c_\alpha(w,w')$.

This is an immediate consequence of 3.3 .

4.3 The following result is due to Jantzen.

PROPOSITION - *Fix* $w \in W_\lambda$, $\alpha \in B_\lambda$. *Then*

$$\text{(i)} \quad c_\alpha(w,w) = \begin{cases} -1 & : \alpha \in \tau_\lambda(w) \\ 1 & : \alpha \notin \tau_\lambda(w) , \end{cases}$$

and for all $w' \in W_\lambda \setminus \{w\}$,

(ii) $c_\alpha(w,w') \in \mathbb{N}$ *and it has the same sign as* $c_\alpha(w,w)$.

For $\alpha \in \tau_\lambda(w)$, the proposition asserts that $a(w)s_\alpha = -a(w)$ which is equivalent to the statement $a(w,w') + a(w,w's_\alpha) = 0$ given in ([6], 2.16 a). For $\alpha \notin \tau_\lambda(w)$, we formally identify w with $\operatorname{ch} M(w\lambda)$ and then $a(w)$ identifies with $\operatorname{ch} L(w\lambda)$. It then suffices to show that there exists $M \in \underline{M}_{\hat{\lambda}}$ which satisfies

$$\operatorname{ch} M = \operatorname{ch} L(w\lambda) + \sum_{w' \in W_\lambda} a(w,w') \operatorname{ch} M(w's_\alpha\lambda) .$$

and $[M : L(w\lambda)] = 2$. The first assertion follows from ([6], 2.18a) the second from ([6], 2.17, Remark 2).

Remark. See also ([12], Sect. 3).

4.4 A subspace of $\mathbb{Q}W_\lambda$ is said to be a-*basal* (resp. a-*convex*) if it is spanned by any subset of $\{a(w) : w \in W_\lambda\}$ (resp. by non-negative integer linear combinations of the $a(w) : w \in W_\lambda$). Given S a subset of $\mathbb{Q}W_\lambda$, let $< S >$ denote the smallest a-basal subspace of $\mathbb{Q}W_\lambda$ containing S.

LEMMA - *Let* S *be an* a-*convex subspace of* $\mathbb{Q}W_\lambda$. *Then for all* $\alpha \in B_\lambda$,

(i) $\mathbb{Q}(s_\alpha S) + \mathbb{Q}S$ *is* a-*convex.*

(ii) $s_\alpha < S > \subset < S > + < s_\alpha S >$

(iii) $< S > s_\alpha \subset < S > + < Ss_\alpha >$.

Fix a vector $v \in S$ which is a non-negative integer linear combination of the $a(w) : w \in W_\lambda$. By 4.1 and 4.2 it follows that $s_\alpha v + v$ is a non-negative integer linear combination of the $a(w) : w \in W_\lambda$ and this establishes (i) .

For (iii) suppose $a(w) \in < v >$. Then by 4.2 , for each $\alpha \in B_\lambda$, either $a(w)s_\alpha = -a(w)$ in which case $a(w)s_\alpha \in < v >$ trivially, or $a(w)s_\alpha$ is a non-negative integer linear combination of the $a(w') : w' \in W_\lambda$. Suppose $a(w') \in < a(w)s_\alpha >$. Then it is enough to show that either $a(w') \in < vs_\alpha >$ or $a(w') \in < v >$ Now the former can only fail to hold through the cancellation of the coefficients of $a(w')$ and by 4.2 this requires the latter to hold. Hence (iii) . Taking 4.1 into account, (ii) obtains from (iii) .

4.5 COROLLARY - *For all* $w \in W_\lambda$, $< \mathbb{Q}W_\lambda a(w) >$ *is a left* W_λ *module.*

By 4.4 (i) , $< \mathbb{Q}W_\lambda a(w) >$ is a a-convex, so the assertion fol lows from 4.4 (ii) .

4.6 PROPOSITION - *Fix* $w,w' \in W_\lambda$. *Suppose* $< \mathbb{Q}W_\lambda a(w') > \subset < \mathbb{Q}W_\lambda a(w) >$. *Then for all* $\alpha \in B_\lambda$,

(i) $a(w')s_\alpha = -a(w')$ *if* $\alpha \in \tau_\lambda(w)$

(ii) $< \mathbb{Q}W_\lambda a(w')s_\alpha > \subset < \mathbb{Q}W_\lambda a(w)s_\alpha >$

By the hypothesis, $a(w') \in < \mathbb{Q}W_\lambda a(w) >$.
If $\alpha \in \tau_\lambda(w)$, then by 4.2 , $a(w)s_\alpha = -a(w)$ and $vs_\alpha = -v$ for each $v \in \mathbb{Q}W_\lambda a(w)$. By 4.4 (i) we can choose v to be a non-

negative integer linear combination of the $a(w'') : w'' \in W_\lambda$ with $a(w') \in < v >$. Taking again account of 4.2 , this gives (i) .

For (ii) we can suppose that $\alpha \notin \tau_\lambda(w)$. By 4.2 , we have $a(w) \in < \mathbb{Q}W_\lambda a(w)s_\alpha >$ and so it is enough to show that $a(w')s_\alpha \in < \mathbb{Q}W_\lambda a(w)s_\alpha > + < \mathbb{Q}W_\lambda a(w) >$. This follows from 4.4 taking $S = \mathbb{Q}W_\lambda a(w)$.

Remark. In general $\mathbb{Q}W_\lambda a(w) \subsetneq < \mathbb{Q}W_\lambda a(w) >$. For example take $B_\lambda = \{\alpha, \beta\}$ of type B_2 (Cartan notation) with $w = s_\alpha s_\beta$. The difficulty that this introduces is overcome by 4.5 and 4.6 .

5 Main Theorems.

5.1 THEOREM - *For all* $w \in W_\lambda$,

(i) $L(M(w_\lambda \lambda) , L(w\lambda)) = V(-w\lambda, -w_\lambda \lambda)$, *up to isomorphism.*

(ii) $r(V(-w\lambda, -w_\lambda \lambda)) = J(\bar{w}\lambda)$.

(i) is a special case of ([10] , 4.7) and (ii) is a special case of ([10] , 4.12) .

5.2 LEMMA - (Notation 2.10) . *For each* $w \in W_\lambda$, $\alpha \in B_\lambda \setminus \tau_\lambda(w)$ *one has*

(i) $\varphi_\alpha \Psi_\alpha (V(-w\lambda, -w_\lambda \lambda)) = L(M(w_\lambda \lambda), \varphi_\alpha \Psi_\alpha L(w\lambda))$, *up to isomorphism.*

(ii) $r(\varphi_\alpha \Psi_\alpha V(-w\lambda, -w_\lambda \lambda)) \supset r(V(-w\lambda, -w_\lambda \lambda))$.

By 5.1 (i), we obtain (ii) as an immediate consequence of 2.10 (ii) . Again by 2.10 (i) , $\varphi_\alpha\Psi_\alpha(V(-w\lambda,-w_\lambda\lambda))$ identifies with a submodule of $L(M(w_\lambda\lambda), \varphi_\alpha\Psi_\alpha L(w\lambda))$. Then to show that they are equal it suffices to apply ([11] , 3.9) again to 5.1 (i) . Here we remark that the analogue of ([11] , 3.9) for the category $\underline{M}$ is also valid and in fact was shown during the course of the proof. Now by ([11] , 3.9 a) ,$\varphi_\alpha\Psi_\alpha L(w\ )$ (resp.

$\varphi_\alpha\Psi_\alpha(V(-w\lambda, -w_\lambda\lambda)))$ admits a unique submodule and a unique quotient and both are isomorphic to $L(w\lambda)$ (resp. $V(-w\lambda,-w_\lambda\lambda))$. Furthermore by ([6] , 2.17, Remark 2 ; [11] , 3.9 b) and ([2] , 7.6.23 ;[3] , I , 4.5) the remaining simple factors of $\varphi_\alpha\Psi_\alpha L(w\lambda)$ (resp.$\varphi_\alpha\Psi_\alpha V(-w\lambda,-w_\lambda\lambda))$ are of the form $L(w'\lambda) : w' \in W_\lambda \setminus \{w\}$ (resp. $V(-w'\lambda,-w_\lambda\lambda) : w' \in W_\lambda \setminus \{w\}$) . Combined with 5.1 (i) this establishes the required assertion.

5.3 For each $I \in \underline{Z}_{\hat{\lambda}}$ we define an a-basal subspace $a(I)$ of $\mathbb{Q}W_\lambda$ by setting

$$a(I) = \oplus \{ \mathbb{Q}a(w) : w \in \underline{V}_\lambda(I) \} .$$

THEOREM - *For each* $I \in \underline{Z}_{\hat{\lambda}}$, $a(I)$ *is a left ideal of* $\mathbb{Q}W_\lambda$.

By 2.7 (i) , (iii) it is enough to prove the assertion for $I \in \underline{X}_{\hat{\lambda}}$. Then by 2.7 (ii) , it suffices to show for all $w \in W_\lambda$, $\alpha \in B \setminus \Psi_\lambda(w)$ that $a(\overline{w'}) \in \langle s_\alpha a(\overline{w}) \rangle$ implies $J(\overline{w'}\lambda) \supset J(\overline{w}\lambda)$. By 4.1 , the first assertion is equivalent to $a(w') \in \langle a(w)s_\alpha \rangle$ which implies that $I(w'\lambda)$ is a simple factor of $\varphi_\alpha \Psi_\alpha L(w_\lambda)$. By 5.1 (i) and 5.2 (i) it follows that $V(-w'\lambda, -w_\lambda\lambda)$ is a simple factor of $\varphi_\alpha \Psi_\alpha V(-w\lambda, -w_\lambda\lambda))$ and then the required assertion follows from 5.1 (ii) and 5.2 (ii) .

5.4 Give W_λ the Duflo order $\subseteq$. Then after Duflo ([4] , Sect. 3 , Cor.1) the map $w \to J(w\lambda)$ of W_λ onto $\underline{X}_{\hat{\lambda}}$ is an order homomorphism. This result also follows from 5.3 taking account of the relation $a(w,w') \neq 0$ if and only if $w \geqslant w'$ (where $\leqslant$ is the Bruhat order on W_λ) . This proof highlights the connection between the order relation on the Verma modules (given by $\leqslant$) and the order relation on the annihilators of their simple quotients (given in part by $\subseteq$) .

5.5 Let Σ_λ denote the set of involutions of W_λ . After Duflo ([4] , Sect. II, 2) one has card $\underline{X}_{\hat{\lambda}} \leqslant$ card Σ_λ . This is also a straightforward consequence of 5.3 .

5.6 Important additional order relations on $\underline{X}_{\hat{\lambda}}$ were given in ([7] , 5.1) . By 4.1 and ([12] , 3.2) these also result from 5.3 .

5.7 The question arises as to whether all the order relations on $\underline{X}_{\hat{\lambda}}$ obtain from 5.3 . This can be put in a number of equivalent forms. First note by 5.3 that we have an injective map $a : I \to a(I)$ of $\underline{Y}_{\hat{\lambda}}$ into the set of all a-basal left ideals of $\mathbb{Q}W_\lambda$.

CONJECTURE A - a *is surjective.*

For each left ideal L of $\mathbb{Q}W_\lambda$ let $\ltimes L \rtimes$ denote the smallest a-basal subspace of the form $a(I) : I \in \underline{Y}_{\hat{\lambda}}$ containing L. Trivially $\langle \mathbb{Q}W_\lambda a(w) \rangle \subset \ltimes \mathbb{Q}W_\lambda a(w) \rtimes$, for all $w \in W_\lambda$.

CONJECTURE B - *For all* $w \in W_\lambda$, *one has*
$\langle \mathbb{Q}W_\lambda a(w) \rangle = \ltimes \mathbb{Q}W_\lambda a(w) \rtimes$.

CONJECTURE C - *For all* $w,w' \in W_\lambda$, $J(w'\lambda) \supset J(w\lambda)$ *if and only if* $a(w') \in \langle \mathbb{Q}W_\lambda a(w) \rangle$.

It is easy to see that all three conjectures are equivalent.

5.8 Let us examine the BJV conditions on $\underline{X}_{\hat{\lambda}}$ referred to in the introduction. These derive from ([1], 2.6) or ([12], 2.3) and translated to the present terminology give the following.

LEMMA - *Fix* $w,w' \in W_\lambda$ *and suppose that* $J(w'\lambda) \supset J(w\lambda)$. *Then*

(i) $\tau_\lambda(w') \supset \tau_\lambda(w)$.

(ii) *For all* $\alpha \notin \tau_\lambda(w)$, *one has*

$$\ltimes \mathbb{Q}W_\lambda a(w')s_\alpha \rtimes \subset \ltimes \mathbb{Q}W_\lambda a(w)s_\alpha \rtimes .$$

(i) is just ([2], 2.14). For (ii) we note that for all $w'' \in W_\lambda$ one has $a(w'') \in \langle a(w)s_\alpha \rangle$ if and only if $L(w''\lambda)$ is a simple factor of $\varphi_\alpha \Psi_\alpha L(w\lambda)$. Hence

$$\sqrt{\operatorname{Ann}(\varphi_\alpha \Psi_\alpha L(w\lambda))} = \cap \{ J(w''\lambda) : a(w'') \in \langle a(w)s_\alpha \rangle \} .$$

Then by 2.7 (iii) we obtain

$$a(\sqrt{\mathrm{Ann}\ \varphi_\alpha \Psi_\alpha L(w_\lambda))}) = \ltimes (QW_\lambda < a(w)s_\alpha >) \rtimes ,$$

$$= \ltimes QW_\lambda a(w)s_\alpha \rtimes , \text{ by } 4.5 \text{ and } 5.3 ,$$

so the required assertion follows from ([12], 2.3).

Remarks. Borho and Jantzen first developed (i) to make a partial separation of the elements of $\underline{X}_{\hat{\lambda}}$. Vogan [12] later noted that combined with (ii) successive applications could lead to a much more refined separation (given sufficient information on the a(w,w')) and Jantzen and later Vogan showed that this was sufficient to obtain a complete separation if B_λ has only type A_n (Cartan notation) factors. Vogan has also shown that complete separation occurs up to rank 4.

5.9 Vogan has expressed the hope that the BJV conditions completely determine the order relations in $\underline{X}_{\hat{\lambda}}$. It is therefore appropriate to point out the following.

THEOREM - *Suppose that* $\underline{X}_{\hat{\lambda}}$ *is determined as an ordered set by* 5.3 *and* 5.8. *Then conjecture* B *holds.*

By 4.6, the BJV conditions are satisfied by taking $\ltimes QW_\lambda a(w) \rtimes \ = \ < QW_\lambda a(w) >$ and by the hypothesis this is the unique solution.

5.10 Fix $w \in W$. Set $\ell(w) = \oplus \{ Qa(w') : J(w\lambda)L(w'\lambda) = 0, w' \in W \}$. It follows from 5.3 and ([7], 4.2; [6], 2.15) that $\ell(w)$ is a left ideal of QW. In particular our analysis can be reformulated without reference to W_λ.

5.11 For each $B' \subset B_\lambda$, set

$$e_{B'} = (\text{card } W_{B'})^{-1} \sum_{w \in W_{B'}} (\det w)w .$$

Then $e_{B'}$ is an idempotent and $s_\alpha e_{B'} = e_{B'}s_\alpha = -e_{B'}$, for all $\alpha \in B'$.

LEMMA - *For all* $B' \subset B$, *one has* $a(J(w_{B'}\lambda)) = \mathbb{Q}W_\lambda e_{B'}$.

Since $w_{B'}\lambda \in \underline{V}(w_{B'}\lambda)$ one has $a(w_{B'}\lambda) \in a(J(w_{B'}\lambda))$. Yet by ([6], 2.23 a)

$$a(w_{B'}\lambda) = \sum_{w' \in W_\lambda} a(w_{B'},w')w'$$

$$= \sum_{w' \in W_{B'}} \det(w'w_{B'})w' .$$

$$= (\text{card } W_{B'})(\det W_{B'})e_{B'} .$$

Hence $e_{B'} \in a(J(w_{B'}\lambda))$. By 5.3, it follows that $\mathbb{Q}W_\lambda e_{B'} \subset a(J(w_{B'}\lambda))$. Yet $\dim \mathbb{Q}W_\lambda e_{B'} = \text{card } (W_\lambda/W_{B'}) = \text{card } \underline{V}_\lambda(w_{B'}\lambda)$ by ([7], 4.2, 4.4) which establishes the assertion of the lemma.

5.12 The above result gives a new interpretation of 5.8 (i). Indeed since $\mathbb{Q}W_\lambda$ is a semisimple Artinian ring, it follows from 5.3 that for each $I \in \underline{Y}_{\hat{\lambda}}$ we can write $a(I) = \mathbb{Q}W_\lambda e(I)$ for some idempotent $e(I)$ of $\mathbb{Q}W_\lambda$. A straightforward computation then gives

COROLLARY - *Fix* $B' \subset B_\lambda$. *Then* $I \supset J(w_{B'}\lambda)$ *if and only if* $e(I)s_\alpha = -e(I)$, *for all* $\alpha \in B'$.

One should also like to interpret $e_{B'}$ as the canonical generator of the corresponding Macdonald representation of W_λ.

REFERENCES.

[1]. W. Borho and J.C. Jantzen, *Über primitive Ideale in der Einhüllenden einer halbeinfacher Lie-algebra*, Invent. Math. 39 (1977) pp. 1 - 53.

[2]. J. Dixmier, *Algèbres enveloppantes*, cahiers scientifiques, XXXVII , Gauthier-Villars, Paris, 1974.

[3]. M. Duflo, *Représentations irréductibles des groupes semi-simples complexes*, Lectures Notes in Mathematics, N° 497, Springer-Verlag, Berlin / Heidelberg / New-York , 1975 , pp. 26 - 88 .

[4]. M. Duflo, *Sur la classification des idéaux primitifs dans l'algèbre enveloppante d'une algèbre de Lie semi-simple*, Ann. Math., 105 1977) pp. 107 - 130.

[5]. J.C. Jantzen, *Zur Charakterformel gewisser Darstellungen halbeinfacher Gruppen und Lie-algebren*, Math. Z., 140 (1974) pp. 127 - 149.

[6]. J.C. Jantzen, *Moduln mit einem höchsten Gewicht*, Habilitationsschrift, Bonn 1977.

[7]. A. Joseph, *A characteristic variety for the primitive spectrum of a semisimple Lie algebra*, preprint, Bonn 1976 (unpublished) . Short version in Lecture Notes in Mathematics, N° 587 , pp. 102 - 118 , Berlin / Heidelberg / New-York 1977.

[8]. A. Joseph, *Towards the Jantzen conjecture*, preprint , Orsay, 1977.

[9]. A. Joseph, *Towards the Jantzen conjecture* II , preprint, Jerusalem, 1978.

[10] A. Joseph, *Dixmier's problem for Verma and principal series submodules*, preprint, Jerusalem, 1978.

[11] D. Vogan, *Irreducible characters of semisimple Lie groups I*, preprint, M.I.T. , 1978.

[12] D. Vogan, *A generalized* τ*-invariant for the primitive spectrum of a semisimple Lie algebra*, preprint, Princeton, 1978.

FUNCTIONS ON THE SHILOV BOUNDARY OF THE GENERALIZED HALF PLANE

M. KASHIWARA and M. VERGNE *
Centre National de la Recherche Scientifique
Massachusetts Institute of Technology

Introduction

Let D be the Siegel upper half plane, i.e.
$D = \{z = x + iy;\ x,y\ n\times n$ symmetric matrices, with y positive definite$\}$.

We consider S the vector space of $n\times n$ symmetric real matrices. We identify S with the Shilov boundary of D by $S = \overline{D} \cap \mathrm{Im}\, z = 0$. The group G of holomorphic transformations of D is the group $Sp(n,\mathbb{R})$. It contains in particular the subgroup P of affine transformations of D ; P is generated by the transformations:

$$g(a)\cdot z = a\, z\, {}^t a \qquad \text{for } a \in GL(n;\mathbb{R})$$

$$t(b)\cdot z = z + b \qquad \text{for } b \in S .$$

The group G is generated by P and the involutive transformation $\sigma(z) = -z^{-1}$ of D . The action of G on D gives rise to an action of G on its Shilov boundary: For $g = \left(\begin{array}{c|c} a & b \\ \hline c & d \end{array}\right)$ a $2n\times 2n$ matrix, belonging to the symplectic group, the corresponding action is $g\cdot x = (ax+b)(cx+d)^{-1}$.

We consider the space $L^2(S)$ of square integrable functions on S . We consider the unitary representation T of G (or of the two-fold covering of G if n is even) on $L^2(S)$ given by:

---

* Supported in part by the National Science Foundation grant number MCS78-02969

$$(T(g)f)(x) = (\det(cx+d))^{-\frac{n+1}{2}} f((ax+b)(cx+d)^{-1})$$

$$\text{for } g^{-1} = \left(\begin{array}{c|c} a & b \\ \hline c & d \end{array}\right) .$$

It is easy to describe the decomposition of $L^2(S)$ with respect to the subgroup P of affine transformations. We identify S with its dual vector space S' by $(x,\xi) = \mathrm{Tr}(x\xi)$. We consider the action $a \mapsto a\,x\,{}^t a$ of $GL(n,\mathbb{R})$ on S, and the corresponding action on S'. Then the space S' breaks under the action of $GL(n;\mathbb{R})$ into a finite number of orbits, classified by the signature. In particular, up to a set of measure zero, $S' = \bigcup_{k_1+k_2=n} \mathcal{O}_{k_1,k_2}$, where $\mathcal{O}_{k_1,k_2}$ is the open set of symmetric matrices of signature (k_1,k_2) with $k_1+k_2 = n$. The Fourier transform $(\mathcal{F}f)(\xi) = \int e^{-i\,\mathrm{Tr}\,\xi x} f(x)\,dx$ is an unitary isomorphism of $L^2(S)$ with $L^2(S')$. Let L_{k_1,k_2} denote the subspace of $L^2(S)$ of functions whose Fourier transform are supported by the set $\overline{\mathcal{O}_{k_1,k_2}}$. Then each L_{k_1,k_2} is invariant under P and $L^2(S) = \oplus L_{k_1,k_2}$ is the decomposition of $L^2(S)$ into irreducible inequivalent representations of P.

Now let us consider the action of the full group G (we suppose n odd for simplicity). Let us consider $H_{n,o}$ the space of functions on S, whose Fourier transform are supported on the convex cone C of positive definite matrices. We can give an alternate description of $H_{n,o}$ as the Hardy space for the domain D, i.e. as a space of boundary values of holomorphic functions on D, via $f(x) = \lim^{L^2}_{y\to 0} F(x+iy)$, with F holomorphic on D . As the action of G extends to an action on holomorphic functions on D , we see that $H_{n,o}$ is invariant by G . Similarly for $H_{o,n}$, as it is a space of boundary values of antiholomorphic functions. (Here we use that n is odd, otherwise $H_{o,n}$ is imbedded naturally in the conjugate representation $\overline{T}$, not isomorphic to T for n even.)

It seems difficult, up to now, to give similarly a description of L_{k_1,k_2} related to the geometry of the complex domain D , hence to conclude some invariance properties of L_{k_1,k_2} under G directly. In fact we have the following beautiful result: Let $p+q = n+1$, with q even, $p,q \neq 0$. We consider $\mathcal{O}_{p-1,q} \cup \mathcal{O}_{p,q-1}$ i.e. for $n = 2m+1$, we arrange the $2(m-1)$ orbits of indefinite matrices (neither positive definite, neither negative) by group of two orbits of adjacent signature. Then the space $H_{p,q} = L_{p-1,q} \oplus L_{p,q-1}$ of functions whose Fourier transform is supported on $\mathcal{O}_{p-1,q} \cup \mathcal{O}_{p,q-1}$ is invariant by G .

This results was initially obtained by the first author, using the calculation of the symbols of the distributions "$(\det x)^s$" , solution of an holonomic system of equations [3]. We present here a simple proof, which uses the Harmonic representation of the group $Sp(n;\mathbb{R})$. The method is close to the one of [2] to prove the unitarity of the solution of the Wave and Maxwell equations, i.e. we use a "flat" description of the cone $\overline{\mathcal{O}_{p-1,q}} \cup \overline{\mathcal{O}_{p,q-1}}$ as the image under the map $\xi \to \xi Q {}^t\xi$ of the vector space $\mathrm{Hom}(\mathbb{R}^{p+q};\mathbb{R}^n)$, for Q a $(n+1)\times(n+1)$ symmetric matrix of signature (p,q) . Then we relate the Fourier transform of the inversion σ , to the Fourier transform on the vector space $\mathbb{R}^{p+q \times n}$.

The analysis of the Weil representation for the pair $SL(2,\mathbb{R}) \times O(p,q)$ by Rallis and Schiffmann [5] (see also R. Howe [1]) suggested to us this application of our method [4].

Let K be the maximal compact subgroup of G , K is isomorphic to $U(n)$. We compute some estimates of the K-types appearing in $H_{p,q}$: If we list the representation of K by their highest weight $(m_1,m_2,\ldots,m_n)$ where (m_i) is a decreasing sequence of n integers, then the highest weights of the representations of K appearing in $H_{p,q}$ are contained in $\mu_o + C_{p,q}$, where μ_o is a character of K , and $C_{p,q}$ denotes the set of highest weights which have less than p positive integers and less than q negative integers.

We will be discussing the philosophy of this striking relation between the analysis of the module $H_{p,q}$ in terms of Fourier series on the compact Lie group K , and in terms of the Euclidean Fourier integral on the vector space S in the article: "K-types and Singular Spectrum" of the same volume.

Our method leads to similar results for other classical domains. If now $D = \{z = x+iy$, with x,y, $n \times n$ hermitian matrices, y positive definite$\}$, the Shilov boundary of D , can be identified to the vector space H of hermitian $(n \times n)$

matrices. We consider the space $L^2(H)$ and the corresponding unitary representations T of the group $G = U(n,n)$ of holomorphic transformations of D. In this case, our result is more easy to describe; Let $p + q = n$ and $H_{p,q}$ be the subspace of functions in $L^2(H)$ whose Fourier transform is supported on the set $\mathcal{O}_{p,q}$ of hermitian matrices of signature (p,q). Then $H_{p,q}$ is invariant by G and $L^2(H) = \oplus H_{p,q}$ is the decomposition of $L^2(H)$ into irreducible unequivalent representations of G.

The interest on this problem has its roots in the joint work of the second author with Hugo Rossi [6] on L^2-functions on the Shilov boundary. We would like to thank him and Eli Stein for many interesting discussions.

I. The Siegel upper half plane and its Shilov boundary.

1. The Domain G/K.

1.1 Let V be a 2n-dimensional real vector space with a non-degenerate skew-symmetric form B. We will consider the group $G = Sp(B)$ of linear transformations of $\mathbb{R}^{2n}$ leaving B invariant.

We write $V = V_1 \oplus V_2$, where $V_1 = \mathbb{R}^n = \sum_{i=1}^{n} \mathbb{R}e_i$, $V_2 = \mathbb{R}^n = \sum_{j=1}^{n} \mathbb{R}f_j$. We write any $2n\times 2n$ matrices $x = \begin{pmatrix} x_{11} & x_{12} \\ x_{21} & x_{22} \end{pmatrix}$ where x_{ij} are $n\times n$ matrices. For $B = \begin{pmatrix} 0 & 1 \\ -1 & 0 \end{pmatrix}$ we have $g \in Sp(B)$ if and only if ${}^t g B g = B$.

1.2 Let $V^{\mathbb{C}}$ be the complexification of V, and extend B to $V^{\mathbb{C}}$. A lagrangian plane λ is an n-dimensional vector subspace of $V^{\mathbb{C}}$ such that $B(\lambda,\lambda) = 0$.

1.3 Let X be the complex manifold of all lagrangian planes in $V^{\mathbb{C}}$. The complexification $G_{\mathbb{C}} = Sp(n,\mathbb{C})$ of G acts on X homogeneously in the obvious way. Let us denote by D the open subset of X consisting of all λ's such that the hermitian form $\frac{1}{i}B(x,\bar{y})$ is positive definite on λ. If $\lambda \in D$, then $\lambda \cap V_2^{\mathbb{C}} = 0$, hence λ can be represented as

$\lambda = \{v + zv;\ v \in V_2^{\mathbb{C}},\ z\colon V_2^{\mathbb{C}} \to V_1^{\mathbb{C}},$ with $z = {}^t z;\ \mathrm{Im}\, z \gg 0\}$. Therefore, D is identified with the Siegel upper half-plane = $\{z,\ n\times n$ complex symmetric matrices, with $\mathrm{Im}\, z \gg 0\}$. The group G acts homogeneously on D by

$$g = \begin{pmatrix} a & b \\ c & d \end{pmatrix}:\ z \to (az+b)(cz+d)^{-1}\ .$$

1.4 The isotropy subgroup K at $z = i$ is given by

$$K = \left\{\begin{pmatrix} a & -b \\ b & a \end{pmatrix}\ ;\ a + ib \in U(n)\right\}$$

For $z, w \in D$, $g \in G$, we have:

$$g\cdot z - \overline{g\cdot w} = (cz+d)^{-1}(z-\bar{w})\ {}^t(c\bar{w}+d)^{-1}\ ,$$

hence the function $\alpha(x) = |\det(x+i)|^2 = \det(1+x^2)$ verifies $\alpha(g\cdot x) = (\det\,(cx+d))^{-2}\,\alpha(x)$ for $g \in K$ and $x \in S(n)$.

Let us consider the vector space $S(n)$ of all $n\times n$ real symmetric matrices $x = {}^t x$. The group $GL(n,\mathbb{R})$ acts on $S(n)$ by $g\cdot x = g\,x\,{}^t g$. We have

1.5 $$\int \varphi(g\cdot x)dx = |\det g|^{-(n+1)} \int \varphi(x)dx\ .$$

$S(n)$ can be considered as the Shilov boundary of the Siegel upper half-plane, via

$$S(n) = \overline{D} \cap \mathrm{Im}\, z = 0\ .$$

The action of G on D gives rise to an action of G on $S(n)$ defined almost everywhere by $g\cdot x = (ax+b)(cx+d)^{-1}$.

1.6 The Jacobian of the transformation

$$x \to (ax+b)(cx+d)^{-1} \quad \text{is} \quad (\det(cx+d))^{-(n+1)} .$$

For $g \in G$ let $g = \begin{pmatrix} a & b \\ c & d \end{pmatrix}$, $j(g,z)$ be an holomorphic function on D such that

$$e^{j(g,z)} = \det(cz+d) \quad \text{for} \quad z \in D .$$

The universal covering group $\widetilde{G}$ of G is defined as the set of elements $(g,j(g,z))$ with the law

$$(g_1,j(g_1,z)) \cdot (g_2,j(g_2,z)) = (g_1g_2,j(g_1,g_2\cdot z) + j(g_2,z)) .$$

For $x \in S(n)$, such that $(cx+d)$ is invertible and $\widetilde{g} = (g,j(g,z))$, we define

$$j(\widetilde{g},x) = \lim_{\substack{z\to x \\ z\in D}} j(g,z) .$$

For $\alpha \in \mathbb{C}$, we denote $e^{\alpha j(\widetilde{g},x)}$ by $(\det(cx+d))^{\alpha}$.

1.7 Consider the unitary representation T of $\widetilde{G}$ in $L^2(S(n))$ defined by

$$(T(\tilde{g}^{-1})f)(x) = (\det(cx+d))^{-(n+1/2)} f((ax+b)(cx+d)^{-1})$$

$$\text{for } \tilde{g} = (\begin{pmatrix} a & b \\ c & d \end{pmatrix}, j(g,z)) \ .$$

This representation gives rise to a unitary representation of the double covering group G_2 (the metaplectic group) of $Sp(n,\mathbb{R})$. When n is odd this is a representation of $Sp(n,\mathbb{R})$ itself.

We will decompose this representation, using the metaplectic representation of the group $Sp(n;\mathbb{R})$.

1.8 Let us consider the subgroup

$$P = \left\{\begin{pmatrix} a & * \\ 0 & {}^t a^{-1} \end{pmatrix}\right\} \text{ of } Sp(n,\mathbb{R}) \ .$$

Each element of P can be written uniquely

$$P = \begin{pmatrix} a & 0 \\ 0 & {}^t a^{-1} \end{pmatrix}\begin{pmatrix} 1 & x \\ 0 & 1 \end{pmatrix}, \text{ with } a \in GL(n;\mathbb{R})$$

and $x \in S(n)$.

The group P acts on $S(n)$ by affine transformations. We first decompose the unitary representation T with respect to P.

Consider the bilinear form $\mathrm{Tr}\, xy$ on $S(n)$ and the Fourier transform

$$\hat{f}(y) = \int e^{-i\,\mathrm{Tr}\,xy} f(x)dx ,$$

then the action of P is given, after Fourier transform, by:

1.9 $$\left(T\begin{pmatrix} a & 0 \\ 0 & {}^t a^{-1}\end{pmatrix}\hat{f}\right)(y) = (\det a)^{n+1/2}\hat{f}(a\,y\,{}^t a)$$

$$\left(T\begin{pmatrix} 1 & x \\ 0 & 1\end{pmatrix}\hat{f}\right)(y) = e^{i\,\mathrm{Tr}\,xy}\,\hat{f}(y) .$$

Let us consider the action of $GL(n,\mathbb{R})$ on $S(n)$ given by $a\cdot y = a\,y\,{}^t a$. Then $S(n)$ is a finite union of orbits under the group $GL(n;\mathbb{R})$ given by the signature of the matrix y . For (k_1,k_2) , two integers with $k_1 + k_2 \leq n$, we denote by $\mathcal{O}_{k_1,k_2}$ the set of matrices of signature (k_1,k_2) .

The orbits $\mathcal{O}_{k_1,k_2}$ for $k_1 + k_2 = n$ are open subsets of $S(n)$ and

$$\overline{\mathcal{O}_{k_1,k_2}} = \bigcup_{\substack{p' \leq k_1 \\ q' \leq k_2}} \mathcal{O}_{(p',q')} .$$

It is clear from the preceding formulas that for $k_1 + k_2 = n$ the subspace $L^2_{k_1,k_2}$ of functions f in $L^2(dx)$ whose Fourier transform is supported in $\overline{\mathcal{O}_{k_1,k_2}}$ is an irreducible invariant subspace of the representation $T|P$. We denote by T_{k_1,k_2} the restriction of $(T|P)$ to the subspace $L^2_{k_1,k_2}$. Hence we have:

1.10 $$T|P = \bigoplus_{k_1+k_2=n} T_{k_1,k_2} \text{ , where each}$$

representation T_{k_1,k_2} is an irreducible representation of P .

Furthermore, from the orbit method, or analyzing the spectrum of the operators $T\begin{pmatrix}1 & x\\ 0 & 1\end{pmatrix}$, we know that $\{T_{k_1,k_2}\}$ are different representations of the group P .

1.11 To simplify the notation, we shall assume that n is odd, i.e. $n+1 = 2m$.

Let us consider the parabolic group $\tilde{P}$ such that

$$\tilde{P} = \{g \in Sp(n,\mathbb{R}) ; g(V_2) = V_2\}$$

i.e. $\tilde{P} = \{g = \begin{pmatrix} a & 0 \\ * & {}^t a^{-1}\end{pmatrix}$, with $a \in GL(n,\mathbb{R})\}$.

We consider the unitary character

$$\epsilon(g) = (\text{sign det } a)^m \quad \text{of } \tilde{P} .$$

Then the unitary induced representation from $\tilde{P}$ to G , by this character, is the representation T .

1.12 More generally, for any integer k , we consider the character $d_k\begin{pmatrix} a & 0 \\ * & {}^t a^{-1}\end{pmatrix} = (\det a)^{-k}$ of the parabolic group $\tilde{P}$.

We consider $\mathcal{C}(k) = \{\varphi,\ C^\infty$ functions on G such that

$$\varphi(g\tilde{p}) = \alpha_k(\tilde{p})^{-1}\varphi(g)\} \quad .$$

The group G acts by left translations on $\mathcal{C}(k)$.

1.13 If $\varphi_1 \in \mathcal{C}(k_1)$, $\varphi_2 \in \mathcal{C}(k_2)$ then $\varphi_1\varphi_2 \in \mathcal{C}(k_1+k_2)$.

On an open set of G , every element of G can be written uniquely as $g = \begin{pmatrix} 1 & x \\ 0 & 1 \end{pmatrix}\begin{pmatrix} a & 0 \\ 0 & {}^t a^{-1} \end{pmatrix}\begin{pmatrix} 1 & 0 \\ y & 1 \end{pmatrix}$ with $x,y \in S(n)$, $a \in GL(n;\mathbb{R})$, hence a function φ of $\mathcal{C}(k)$ is determined by $f(x) = \varphi\begin{pmatrix} 1 & x \\ 0 & 1 \end{pmatrix}$ which is a C^∞ function on $S(n)$. We denote by $\overline{\mathcal{C}}(k)$ the space of C^∞ functions on $S(n)$ obtained by restriction of a function $\varphi \in \mathcal{C}(k)$. The left action of G in $\mathcal{C}(k)$ becomes on $\overline{\mathcal{C}}(k)$

$$(T_k(g)f)(x) = (\det(cx+d))^{-k}f((ax+b)(cx+d)^{-1}) \quad \text{for} \quad g^{-1} = \begin{pmatrix} a & b \\ c & d \end{pmatrix}$$

The formula for the action of $T_k(g)$ is not defined everywhere. However, for $f \in \overline{\mathcal{C}}(k)$, $T_k(g)f$ extends to an analytic function on $S(n)$. Reciprocally if f is such that $T_k(g)f$ can be extended as an analytic function on $S(n)$ for every $g \in G$, then f is the restriction of a function φ of $\mathcal{C}(k)$ defined by $\varphi(g) = (T_k(g^{-1})f)(0)$.

For $\varphi \in \mathcal{C}(k)$ we define $\|\varphi\|^2 = \int_K |\varphi(g)|^2 dg$. This is a norm on $\mathcal{C}(k)$ invariant by the left action of K . For this norm the action of $g \in G$ becomes a bounded transformation. It follows from 1.4, 1.6 that for $\varphi \in \mathcal{C}(k)$

and $f(x) = \varphi\begin{pmatrix} 1 & x \\ 0 & 1 \end{pmatrix}$ that

$$\int |f(x)|^2 \det (1+x^2)^{-(\frac{n+1}{2} - k)} dx = ||\varphi||^2 < \infty .$$

In particular for $k = \frac{n+1}{2}$ the restriction of any function $\varphi \in \mathcal{C}(\frac{n+1}{2})$ is in $L^2(dx)$.

(For k being a half integer, we define similarly the representations T_k for the metaplectic group.)

2. The Harmonic representation of $Sp(n,\mathbb{R})$.

We define the following elements of $Sp(n;\mathbb{R})$

$$g(a) = \begin{pmatrix} a & 0 \\ 0 & ({}^t a)^{-1} \end{pmatrix} \quad \text{for } a \in GL(n,\mathbb{R})$$

$$t(b) = \begin{pmatrix} 1 & b \\ 0 & 1 \end{pmatrix} \quad \text{for } b \in S(n)$$

$$\sigma = \begin{pmatrix} 0 & 1 \\ -1 & 0 \end{pmatrix} .$$

These elements generate $Sp(n;\mathbb{R})$.

The following choices of $L(g)$ determine a representation of G_2 on $L^2(\mathbb{R}^n)$:

$$(L(\dot{g}(a))f)(\xi) = (\det a)^{\frac{1}{2}} f({}^t a \cdot \xi)$$

$$(L(t(b))f)(\xi) = e^{-\frac{i}{2}(b\xi,\xi)} f(\xi)$$

$$(L(\dot{\sigma})f)(\xi) = \left(\frac{i}{2\pi}\right)^{n/2} \int_{\mathbb{R}^n} e^{i(\xi,\xi')} f(\xi')d\xi'$$

(i.e. $L(\sigma)$ is proportional to the Fourier transform). The precise definitions of $\dot{g}(a) \in G_2$, $\dot{\sigma} \in G_2$, $(\det a)^{\frac{1}{2}}$, $i^{\frac{1}{2}}$ are as in [4].

We consider the representation $\overline{L}$ of G_2 in $L^2(\mathbb{R}^n)$ given by $\overline{L}f = \overline{L\overline{f}}$, and the representation $L_{p,q} = \overset{p}{\otimes} L \otimes \overset{q}{\otimes} \overline{L}$ of G_2 in $L^2(\mathbb{R}^{n\times(p+q)})$.

Let $M_{n,p+q} = \mathrm{Hom}_{\mathbb{R}}(\mathbb{R}^{p+q}, \mathbb{R}^n)$. We write an element $\xi: \mathbb{R}^{p+q} \to \mathbb{R}^n$ as $\xi = (\alpha, \beta)$ where $\alpha: \mathbb{R}^p \to \mathbb{R}^n$, $\beta: \mathbb{R}^q \to \mathbb{R}^n$. We consider the canonical form Q on $\mathbb{R}^{p+q}$ of signature (p,q) , then $\xi Q {}^t\xi$ is a $n \times n$ symmetric matrix which is equal to $\alpha {}^t\alpha - \beta {}^t\beta$.

$L_{p,q}$ can be realized in $L^2(M_{n,p+q})$ as follows:

$$(L_{p,q}(g(a))f)(\xi) = (\det a)^{\frac{p+q}{2}} f({}^t a \xi)$$

$$(L_{p,q}(t(b))f)(\xi) = e^{-\frac{i}{2}\mathrm{Tr}(\xi Q {}^t\xi b)} f(\xi)$$

$$(L_{p,q}(\sigma)f)(\xi) = \left(\frac{i}{2\pi}\right)^{\frac{(p-q)n}{2}} \int_{M_{n,p+q}} e^{i \,\mathrm{Tr}\, \xi Q {}^t\xi'} f(\xi')d\xi' .$$

2.2 On $L^2(M_{n,p+q})$ the action of the group $O(p,q)$ given by $(h\cdot f)(\xi) = f(\xi h)$ commutes with the representation $L_{p,q}$ of G_2 .

In this paper we will consider the isotypic component corresponding to the identity representation of $O(p,q)$.

2.3 We consider p and q two integers with $p+q = k$ and q even. We then have that for $\tilde{g} \in G_2$,

$$(\det(cx+d))^{p/2} \overline{\det(cz+d)^{q/2}} = (\det(cx+d))^{k/2}$$

as in this case $(\det(cx+d))^{q/2}$ is real.

We recall that for $q = 0$ [4] the operator

$$(\mathcal{I}_{p,o}\varphi)(x) = \int_{M_{n,p}} e^{\frac{i}{2}\mathrm{Tr}\,(\alpha {}^t\alpha)x} \varphi(\alpha)d\alpha$$

intertwines the representation $L_{p,o}$ with the representation

$$(T_{p/2}(g)f)(x) = (\det(cx+d))^{-p/2} f((ax+b)(cx+d)^{-1})$$

of G_2 , $g^{-1} = \begin{pmatrix} a & b \\ c & d \end{pmatrix}$. If $\varphi \in \mathcal{S}$, the Swartz space on $M_{n,p}$, $\mathcal{J}_{p,o}\varphi \in \mathcal{C}(p/2)$:

Similarly if we consider the representation $L_{o,q} = \overline{L_{q,o}}$ the operator

$$\mathcal{J}_{o,q} = \overline{\mathcal{J}_{q,o}} = \int_{M_{n,q}} e^{-\frac{i}{2}\operatorname{Tr}(\beta\, {}^t\beta)x} \varphi(\beta)d\beta$$

intertwines the representation $L_{o,q}$ with the representation

$$(\overline{T}_{q/2}f)(x) = (\det(cx+d))^{-q/2} f((ax+b)(cx+d)^{-1})$$

For q even $\overline{T}_{q/2} = T_{q/2}$.

If $\varphi \in \mathcal{S}$, then $\mathcal{J}_{o,q}\varphi \in \overline{\mathcal{C}}(q/2)$. We consider the map (1.13) from $\mathcal{C}(p/2) \otimes \overline{\mathcal{C}}(q/2) \to \mathcal{C}(\frac{p+q}{2})$ given by $(f_1 \otimes f_2)(x) = f_1(x)f_2(x)$. The composed operator $m \cdot (\mathcal{J}_{p,o} \otimes \mathcal{J}_{o,q}) = \mathcal{J}_{p,q}$ will be an intertwining operator between the representation $L_{p,q}$ and $T_{p+q/2}$. It is given on $\varphi = \varphi_1 \otimes \varphi_2$ $(\varphi_1 \in \mathcal{S}(\alpha),\ \varphi_2 \in \mathcal{S}(\beta))$, by

$$(\mathcal{J}_{p,q}\varphi)(x) = \int e^{\frac{i}{2}\operatorname{Tr}(\alpha\, {}^t\alpha - \beta\, {}^t\beta)x} \varphi_1(\alpha)\varphi_2(\beta)d\alpha d\beta \quad .$$

This leads to the following definition: Let us consider the Schwartz space $\mathcal{S}(M_{n,p+q})$ of rapidly decreasing functions on the vector space $M_{n,p+q}$. It is clear that $\mathcal{S}$ is stable by the action of $L_{p,q}$. We consider the operator:

$$(\mathcal{I}_{p,q}\varphi)(x) = \int_{M_{n,p+q}} e^{\frac{i}{2}\mathrm{Tr}(\xi Q\, {}^t\xi)x} \varphi(\xi)d\xi$$

for $\varphi \in \mathcal{S}$.

2.4 <u>Proposition</u>: $\mathcal{I}_{p,q}$ sends $\mathcal{S}(M_{n,p+q})$ in $L^2(S(n), \det(1+x^2)^{k-(n+1)/2}dx)$ and intertwines the representation $L_{p,q}$ with $T_{k/2}$.

<u>Proof</u>: It is clear that for $\varphi \in \mathcal{S}(M_{n,p+q})$ the function $\mathcal{I}_{p,q}\varphi$ is defined and is an analytic function of x .

We check that $\mathcal{I}_{p,q}$ is an intertwining operator as in [4]. The necessary commutations relations are obvious to check on the elements $t(b)$, $g(a)$. We check the action of $\sigma = \begin{pmatrix} 0 & 1 \\ -1 & 0 \end{pmatrix}$. As $L_{p,q}(\sigma)f$ is a multiple the Fourier transform $\hat{f}$ of f with respect to the bilinear form $\xi Q\, {}^t\xi'$, we have

$$(\mathcal{I}L_{p,q}(\sigma)f)(x) = \text{const.} \int e^{\frac{i}{2}\mathrm{Tr}(\xi Q\, {}^t\xi)x} \hat{f}(\xi)d\xi$$

$$= \text{const.} \int (e^{\frac{i}{2}\mathrm{Tr}(\xi Q\, {}^t\xi)x})^{\wedge} f(\xi)d\xi$$

by the Plancherel formula.

But it is easy to check (using homogeneity properties with respect to $GL(n,\mathbb{R})$) that

$$\mathcal{F}(e^{\frac{i}{2}\mathrm{Tr}(\xi Q\,{}^t\xi)x}) = (\det x)^{-k/2}\, e^{-\frac{i}{2}\mathrm{Tr}(\xi Q\,{}^t\xi)x^{-1}}$$

in the distribution sense. Hence we obtain the necessary commutation relations.

As $\mathcal{S}(M_{n,p+q})$ is stable by the action of $L_{p,q}$ this implies that for every $g \in G$ and $f \in \mathcal{S}$ and function $T_{k/2}(g)\mathcal{F}_{p,q}f(x)$ is analytic in x (being $\mathcal{F}_{p,q}(L_{p,q}(g)f)$, with $L_{p,q}(g)f \in \mathcal{S}$). Hence for $f \in \mathcal{S}$, $(\mathcal{F}_{p,q}f) \in \overline{\mathcal{C}}(p+q/2)$, in particular is in $L^2(S(n), \det(1+x^2)^{\frac{k-(n+1)}{2}}\, dx)$.

2.5 Let us study the image of $\mathcal{F}_{p,q}(\mathcal{S})$. We consider the map $\xi \mapsto \xi Q\,{}^t\xi$ of $M_{n,p+q}$ into $S(n)$.

2.6 <u>Lemma</u>. The image under the map $\xi \mapsto \xi Q\,{}^t\xi$ of $M_{n,p+q}$ consists of all symmetric matrices of signature (p',q') with $p' \leq p$, $q' \leq q$ (and $p'+q' \leq n$).

<u>Proof</u>: Let $W \subset \mathbb{R}^n$ a vector subspace such that the restriction of the form $\xi Q\,{}^t\xi = \alpha\,{}^t\alpha - \beta\,{}^t\beta$ is positive definite on W. We want to prove that $\dim W \leq p$.

We have for every $w \in W$, $w \neq 0$

$$\langle(\alpha\,{}^t\alpha - \beta\,{}^t\beta)w, w\rangle > 0, \text{ so}$$

$$\langle{}^t\alpha w, {}^t\alpha w\rangle > \langle{}^t\beta w, {}^t\beta w\rangle \geq 0 .$$

So ${}^t a w \neq 0$. Hence the map ${}^t a = W \to \mathbb{R}^p$ is injective and $\dim W \leq p$. Similarly we see that if $W' \subset \mathbb{R}^n$ is such that the restriction of the form $\xi Q {}^t\xi$ is negative definite then $\dim W' \leq q$. This proves the lemma.

2.7 We denote by $S_{p,q} \subset S(n)$ the set of symmetric matrices of signature (p',q') with $p' \leq p$ and $q' \leq q$. For $p' + q' = n$ we recall that $\mathcal{O}_{p',q'}$, the set of symmetric matrices of signature (p',q') , is an open orbit under the action of $GL(n,\mathbb{R})$ in $S(n)$. We denote by $\mathcal{O}^+$ the set of positive definite matrices, $\mathcal{O}^-$ the set of negative definite matrices.

For $p + q < n$, $S_{p,q}$ is contained in the closed subset of $S(n)$ of matrices of rank $\leq n$. In particular $S_{p,q}$ is of measure zero in $S(n)$. For $p + q \geq n$, then $S_{p,q}$ is the adherence of the open orbits $\mathcal{O}_{p',q'}$ with $p' \leq p$, $q' \leq q$, $p' + q' = n$ (for $p \geq n$ and $q \geq n$, then $S_{p,q} = S(n)$, i.e. we have no restriction on the support).
For $p + q = n + 1$, we have:

$$S_{p,q} = \mathcal{O}_{p-1,q} \cup \mathcal{O}_{p,q-1}$$

up to a set of measure zero, whenever $p \geq 1$, $q \geq 1$

$$S_{p,o} = \overline{\mathcal{O}^+}$$

$$S_{o,q} = \overline{\mathcal{O}^-} \ .$$

When (p,q) varies over the integers (p,q) with $p+q=n+1$, q even; then we have that $S(n)$ up to a set of measure zero is the disjoint union of the set $S_{p,q}$.

2.8 Let us suppose that $p+q \geq n$. Let dy be the Euclidean measure on $S(n)$. We consider dy as a measure on $S_{p,q}$. There exists measures $d_y\xi$ on the varieties $\frac{\xi Q\, {}^t\xi}{2} = y$ such that, for any function φ on $M_{n,p+q}$ compactly supported, we have

$$\int \varphi(\xi)d\xi = \int\left(\int_{\xi Q\, {}^t\xi=y} \varphi(\xi)d_y\xi\right)dy .$$

Let $\bar{\varphi}(y) = \int_{\xi Q\, {}^t\xi=y} \varphi(\xi)d_y\xi$. $\bar{\varphi}(y)$ is a function on $S(n)$ supported on $S_{p,q}$. We can write $(\mathcal{F}_{p,q}\varphi)(x) = \int e^{i\,\mathrm{Tr}\, yx}\bar{\varphi}(y)dy$, hence the function $(\mathcal{F}_{p,q}\varphi)(x)$ is the Fourier transform of a function supported on $S_{p,q}$.

2.9 Let us consider the subspace $\mathcal{F}_{p,q}(\mathcal{S})$ of $L^2(\det(1+x^2)^{\frac{k-(n+1)}{2}}dx)$ and its closure $H_{p,q}$ in L^2. Then $H_{p,q}$ consists of functions $f(x)$ in $L^2(S(n),\det(1+x^2)^{\frac{k-(n+1)}{2}}dx)$ whose Fourier transform, as a distribution, has support in $S_{p,q} = \bigcup_{\substack{p'\leq p\\ q'\leq q}} \mathcal{O}_{p',q'}$.

2.10 In particular, for the values (p,q) with q even

$p + q = n + 1$, we conclude that $H_{p,q}$ is an invariant subset under G of functions in $L^2(dx)$ whose Fourier transform is supported in $S_{p,q}$. $H_{p,q}$ being invariant under P, we must have (for $p \geq 1$, $q \geq 1$)

$$H_{p,q} = L^2_{p-1,q} \oplus L^2_{p,q-1}$$

i.e. $H_{p,q}$ is the space of function f in $L^2(dx)$ whose Fourier transform is supported on $S_{p,q}$.

The space $H_{p,q}$ is invariant under G. We will prove now that it is an irreducible subspace of G. If not $H_{p,q} = L_{p-1,q} \oplus L_{p,q-1}$ will be the decomposition of $H_{p,q}$ under G.

Let us consider the function $\alpha_{p,q} = \det(x+i)^{-p/2}\det(x-i)^{-q/2}$. It is easy to check from 1.4 that $\alpha_{p,q}$ transforms under a character χ of the group K.

We have [4]

$$\det(x+i)^{-p/2} = \int_{M_{n,p}} e^{\frac{i}{2}\mathrm{Tr}(\alpha\, {}^t\alpha)x}\, e^{-\frac{\mathrm{Tr}(\alpha\, {}^t\alpha)}{2}}\, d\alpha ,$$

hence

$$\det(x+i)^{-p/2}\det(x-i)^{-q/2} = \int_{M_{n,p+q}} e^{\frac{i}{2}\mathrm{Tr}(\xi Q\, {}^t\xi)x}\, e^{-\mathrm{Tr}\frac{(\xi\, {}^t\xi)}{2}}\, d\xi$$

The function $\xi \to e^{-\frac{\mathrm{Tr}(\xi\, {}^t\xi)}{2}}$ is in $\mathcal{S}(M_{n,p+q})$ so $\alpha_{p,q} \in H_{p,q}$. Furthermore this function of ξ is strictly positive on $M_{n,p+q}$ hence the support of $\bar{\varphi}(y)$ is the entire set

$S_{p,q} = \overline{\mathcal{O}_{p,q-1}} \cup \overline{\mathcal{O}_{p-1,q}}$.

The multiplicity of χ in $C(k/2)$ is one. Hence if $H_{p,q}$ was reducible, then $\alpha_{p,q}$ will belong to either $L_{p,q-1}$ or $L_{p-1,q}$. This proves the

2.11 Theorem: The representation T of G_2 in $L^2(S(n))$ is the sum of the irreducible representations $T_{p,q}$ (for $p+q = n+1$, q even) in $H_{p,q} = L_{p-1,q} \oplus L_{p,q-1}$.

Hence we see that we obtain the decomposition of T by considering the Fourier transforms of functions supported on the union of two orbits of $GL(n;\mathbb{R})$ in $S(n)$, i.e. matrices with signatures $(p-1,q)$ or $(p,q-1)$.

3. The K-types of the representation $T_{p,q}$.

We will suppose, just in order to simplify the notations that p and q are even, and we set $p + q = k$.

Let us consider the space $\mathcal{S}(M_{n,p+q})$ and its dual $\mathcal{S}'(M_{n,p+q})$. They are both stable under the action of $G \times O(p,q)$. We can form for $\varphi \in \mathcal{S}$, $\varphi' \in \mathcal{S}'$ the function $(L_{p,q}(g)^{-1}\varphi,\varphi') = (\varphi, L_{p,q}(g)\varphi')$ which is a C^∞ function on G .

Let us consider the function $\varphi_0 = 1$ on $M_{n,p+q}$. We consider φ_0 as an element of $\mathcal{S}'$, by $(\varphi_0,\varphi) = \int \varphi(\xi)d\xi$. Obviously φ_0 is invariant under the action of $O(p,q)$.

Let us look at the action of the group G on φ_0 . We have $L_{p,q}(g(a))f_0 = (\det a)^{p+q/2}f_0$.

Let $\mathfrak{g}$ be the Lie algebra of G . We write $\mathfrak{g} = \mathfrak{g}_{-1} \oplus \mathfrak{g}_0 \oplus \mathfrak{g}_1$ with

$$\mathfrak{g}_1 = \{\begin{pmatrix} 0 & x \\ 0 & 0 \end{pmatrix}; x \in S(n)\}$$

$$\mathfrak{g}_{-1} = \{\begin{pmatrix} 0 & 0 \\ y & 0 \end{pmatrix}; y \in S(n)\} = \sigma(\mathfrak{g}_1) \quad .$$

$\mathfrak{g}_1$ acts on $\mathcal{S}$ by multiplications operators. $\mathfrak{g}_{-1}$ atcs on $\mathcal{S}$ by derivations. Hence φ_0 is annihilated by $\mathfrak{g}_{-1}$. So φ_0 transform under a one-dimensional representation of the parabolic group $\tilde{P}$.

It follows that for $\varphi \in \mathcal{S}$ the function $I(\varphi)(g) = (\varphi, L_{p,q}(g)\varphi_0)$ is $\mathcal{C}(k/2)$; moreover $I(\varphi)$ intertwines the

representation $L_{p,q}$ of G in $\mathcal{A}$ with left translations.
We have

$$(I\varphi)\begin{pmatrix}1 & x\\ 0 & 1\end{pmatrix} = (\varphi, L_{p,q}(t(x))\varphi_0)$$
$$= (L_{p,q}(t(x)^{-1})\varphi, \varphi_0)$$
$$= \int e^{\frac{i}{2}\xi Q\, {}^t\xi x} \varphi(\xi)d\xi = (\mathcal{J}_{p,q}\varphi)(x) .$$

We will obtain some information on the K-types of the representation $T_{p,q}$ of G in $H_{p,q}$ by analyzing the K-types of the module generated by φ_0.

We consider $\mathcal{A}'(0)$ the space of tempered distributions on $M_{n,p+q}$ invariant under $O(p,q)$. $\mathcal{A}'(0)$ is stable under G.

For $\tau \in K^\wedge$ we consider $\mathcal{A}'(0,\tau)$ the subspace of $\mathcal{A}'(0)$ of distributions transforming under K according to the representation τ. We denote τ' the contragredient representation to τ.

3.1 We denote by $P(p,q)$ the set of $\tau \in K^\wedge$ such that $\mathcal{A}'(0,\tau) \neq \{0\}$.

We consider the Hilbert space $H_{p,q}$ and for $\tau \in \hat{K}$, we denote by $H_{p,q}(\tau)$ the isotypic component of type τ.

3.2 Lemma. If $H_{p,q}(\tau) \neq 0$, then $\tau' \in P(p,q)$.

<u>Proof</u>: The projector of $H_{p,q}$ in $H_{p,q}(\tau)$ is given by $P_\tau v = \int_K \overline{\operatorname{Tr}\tau(k)}\,\tau(k)v\,dk$. Hence $H_{p,q}(\tau) \neq 0$ if and only if $P_\tau \neq 0$. By definition of $H_{p,q}$ there exist $\varphi \in \mathcal{J}$ such that $P_\tau(I\varphi) \neq 0$. But

$$\begin{aligned}(P_\tau(I\varphi))(g) &= \int_K \overline{\operatorname{Tr}\tau(k)}\,(I\varphi)(k^{-1}g)dk \\ &= \int_K \overline{\operatorname{Tr}\tau(k)}\,(L_{p,q}(g^{-1})L_{p,q}(k)\varphi,\varphi_0)dk \\ &= (\varphi, \int_K \overline{\operatorname{Tr}\tau(k)}\,L_{p,q}(k)^{-1}(L_{p,q}(g)\varphi_0)dk)\end{aligned}$$

But as $\varphi_0 \in \mathcal{J}'(0)$, $L_{p,q}(g)\varphi_0 = \lambda \in \mathcal{J}'(0)$ and

$$\int_K \overline{\operatorname{Tr}\tau(k)}(L_{p,q}(k^{-1})\lambda)dk = \int_K \overline{\operatorname{Tr}\,{}^t\tau(k^{-1})}(L_{p,q}(k)\lambda)\,dk$$

$$\in \mathcal{J}'(0,\tau') \text{ and is } \neq 0 .$$

This proves the lemma.

We will now analyze the set $P(p,q)$. For this, it will be more convenient to use the Bargmann model for the representation of the Harmonic representation. We will describe this model. Let $\mathbb{C}^n$ be equipped with the canonical hermitian inner product $\langle u,u'\rangle$. We consider

$$\mathcal{H} = \{f \text{ antiholomorphic functions on } \mathbb{C}^n \text{ such that } \int |f(u)|^2 e^{-\frac{\langle u,u\rangle}{2}}\,du\,d\bar{u} < \infty .$$

We consider on C^n the non-degenerate bilinear form $B(u,v) = -\mathrm{Im}\langle u,v\rangle$. Let us form the $(2n+1)$ dimensional Heisenberg Lie algebra $\mathfrak{n} = C^n \oplus \mathbb{R}E$, with the law $[u,v] = B(u,v)E$. ($\mathbb{R}E$ being the center of $\mathfrak{n}$.)

We consider the Heisenberg group N of Lie algebra $\mathfrak{n}$. The operator

$$(W(\exp u)f)(z) = e^{-\frac{\langle u,u\rangle}{4}} e^{\frac{\langle u,z\rangle}{2}} f(z-u)$$

are unitary operators on $\mathcal{H}$ and verifies the relation

$$W(\exp u_0)W(\exp u_1) = e^{\frac{i}{2}\mathrm{Im}\langle u_0,u_1\rangle} W(\exp(u_0+u_1)) .$$

We can consider W as an unitary representation of N if we set $W(\exp tE) = e^{-it}\mathrm{Id}_{\mathcal{H}}$.

We consider now $C^n = \mathbb{R}^n + i\mathbb{R}^n = V_1 \oplus V_2 = V$, and the representation U of the Heisenberg group N in $L^2(V_2)$ given by:

$$(U(\exp y_1)f)(x) = e^{-i(y_1,x)} f(x)$$

$$(U(\exp y_2)f)(x) = f(x-y_2)$$

$$(U(\exp tE)f)(x) = e^{-it} f(x) .$$

By definition, the harmonic representation $\overline{L}$ is the unique representation of G_2 in $L^2(V_2)$ verifying

$$\overline{L}(g)U(\exp v)\overline{L}(g)^{-1} = U(\exp g\cdot v) .$$

We consider the transformation

$$(A\varphi)(z) = \int e^{\frac{1}{4}\langle\bar{z},z\rangle} e^{+i\langle\xi,z\rangle} e^{-\frac{1}{2}\langle\xi,\xi\rangle} \varphi(\xi)d\xi .$$

We have the

3.4 Lemma: A is a unitary isomorphism between $L^2(V_2)$ and $\mathcal{H}$ such that

$$A \bullet U(n) = W(n) \bullet A \quad \text{for } n \in N .$$

We remark that we can extend A to $\mathcal{S}'(\mathbb{R}^n)$ by

$$(A'\lambda)(z) = \int e^{\frac{1}{4}\langle z,\bar{z}\rangle} e^{-i\langle z,\xi\rangle} e^{-\frac{1}{2}\langle\xi,\xi\rangle} d\lambda(\xi) .$$

$(A'\lambda)(z)$ is an analytic function on $\mathbb{C}^n$. For $\varphi \in \mathcal{S}(\mathbb{R}^n)$, $\lambda \in \mathcal{S}'(\mathbb{R}^n)$, we have

$$\langle\lambda,\varphi\rangle = \int (A\varphi)(z)A'(\lambda)(z)\, e^{-\frac{\langle z,z\rangle}{2}}\, dz\, d\bar{z} .$$

For example, the image of the function 1 will be given by

$$e^{-\frac{1}{4}z^2} \quad (\text{where } z^2 = \langle z,\bar{z}\rangle = z_1^2 + \dots + z_n^2) .$$

The group $U(n)$ acts on $\mathbb{C}^n$ preserving the hermitian form

$\langle u,v\rangle$, a fortiori B . This gives a map $U(n) \to G = Sp(n,\mathbb{R})$ which coincides with the identification of $U(n)$ with K given by 1.4. For $g \in U(n)$ we consider the action $(U_1(g)f)(x) = f(g^{-1}x)$ in $\mathcal{H}$; $U_1(g)$ is an unitary operator on $\mathcal{H}$ verifying

$$U_1(g)W(\exp v)U_1(g)^{-1} = W(\exp g\cdot v) .$$

Let us consider the representation Λ of G_2 in $\mathcal{H}$ transported from the representation L of G_2 in $L^2(V_2)$ via the isomorphism A . We have for $g \in U(n)$ the formula:

3.5 $$(\Lambda(g)f)(u) = (\det g)^{\frac{1}{2}}f(g^{-1}\cdot u) .$$

Similarly we consider the space $\overline{\mathcal{H}} = \{f$ holomorphic on $\mathbb{C}^n$; $\int |f(v)|^2 e^{-\langle v,v\rangle/2}\, dv\, d\bar{v} < \infty\}$, the isomorphism $\overline{A}$: $L^2(V_2) \to \overline{\mathcal{H}}$ given by:

$$(\overline{A}\varphi)(z) = \int e^{\frac{1}{4}\langle z,\bar{z}\rangle} e^{-i\langle z,\xi\rangle} e^{-\frac{1}{2}\langle \xi,\xi\rangle} \varphi(\xi)d\xi$$

and the representation $\overline{\Lambda}$ of G_2 in $\overline{\mathcal{H}}$ given by $\overline{\Lambda} = A \circ \overline{L} \circ A^{-1}$ then we have

3.6 $$(\overline{\Lambda}(g)f)(v) = (\det g)^{-\frac{1}{2}}f(g^{-1}v) \quad \text{for } g \in U(n) .$$

Finally let us consider the representation $L_{p,q}$ of G_2 in $L^2(V_2)$. We introduce the space $M_{n,p}(\mathbb{C}) \times M_{n,q}(\mathbb{C}) = M_{n,p,q}(\mathbb{C}) = \mathrm{Hom}_{\mathbb{C}}(\mathbb{C}^p,\mathbb{C}^n) \times \mathrm{Hom}_{\mathbb{C}}(\mathbb{C}^q;\mathbb{C}^n)$. We denote by (u,v) an element of $M_{n,p,q}(\mathbb{C})$. We define the space $\mathcal{H}_{p,q} = \{f(u,v)$ antiholomorphic in u , holomorphic in v such that

$$\int |f(u,v)|^2 e^{-\operatorname{Tr}\frac{uu^*+vv^*}{2}} du\, d\bar{u}\, dv\, d\bar{v} < \infty\}$$

the isomorphism:

$A_{p,q}$: $L^2(M_{n,p+q})$ in $\mathcal{H}_{p,q}$ given by

$$(A_{p,q}\varphi) = \int e^{\frac{1}{4}\operatorname{Tr}(\bar{u}\,{}^t\bar{u})} e^{\frac{1}{4}\operatorname{Tr}(v\,{}^tv)} e^{i\operatorname{Tr}\bar{u}\,{}^t\alpha} e^{-i\operatorname{Tr}v\,{}^t\beta}$$

$$e^{-\frac{1}{2}\operatorname{Tr}(\alpha\,{}^t\alpha + \beta\,{}^t\beta)} \varphi(\alpha,\beta)\,d\alpha\, d\beta$$

and the representation $\Lambda_{p,q} = A_{p,q} \bullet L_{p,q} \bullet A_{p,q}^{-1}$.
Then for $g \in U(n)$:

$$(\Lambda_{p,q}(g)f)(u,v) = (\det g)^{\frac{p-q}{2}} f(g^{-1}u, g^{-1}v) .$$

Let $\mathcal{O}_{p,q}$ be the space of functions on $M_{n,p,q}(\mathbb{C})$ holomorphic in u , antiholomorphic in v . We have an injection $A'_{p,q}$ of $\mathcal{S}'(M_{n,p+q}(\mathbb{R}))$ in $\mathcal{O}_{p,q}$ such that for $\varphi \in \mathcal{S}$, $\lambda \in \mathcal{S}'$

$$\int (A_{p,q}\varphi)(u,v)(A'_{p,q}\lambda)(u,v)e^{-\frac{1}{2}\operatorname{Tr}(uu^*+vv^*)} du\, dv = (\varphi,\lambda) .$$

3.7 We consider the action of $U(n)$ in $\mathcal{O}(p,q)$ given by $(g\cdot f)(u,v) = (\det g)^{-(p-q)/2} f(g^{-1}u, g^{-1}v)$. Then $\mathcal{S}'$ is identified via $A'_{p,q}$ to a submodule of $\mathcal{O}_{p,q}$ under this action of $U(n)$.

Now we will identify the action of $O(p,q)$ on $\mathcal{H}_{p,q}$. We consider the action of $O(p,q)$ on $\mathbb{R}^{p+q}$. A function $\varphi(\alpha_1,\alpha_2,\ldots,\alpha_p,\beta_1,\beta_2,\ldots,\beta_q)$ on $\mathbb{R}^{p+q}$ is invariant under $O(p,q)$, if

α) φ is invariant under the action of $O(p) \times O(q)$,

β) $\alpha_i \frac{\partial}{\partial \beta_j} + \beta_j \frac{\partial}{\partial \alpha_j} \cdot \varphi = 0$, for every i,j.

Let us consider the space $M_{n,p,q}(\mathbb{C})$. The group $O(p) \times O(q)$ acts on $M_{n,p,q}(\mathbb{C})$ by $(u,v) \to (u\sigma_1, v\sigma_2)$ for $(\sigma_1,\sigma_2) \in O(p) \times O(q)$.

For u_i a column vector in the matrix u (i.e. $u_i \in \mathbb{C}^n$) v_j a column vector in the matrix v, we denote by $v_j\bar{u}_i$ the function $\sum_{\alpha=1}^{n} v_j^{\alpha}\bar{u}_i^{\alpha}$ and by $\frac{\partial}{\partial v_j}\frac{\partial}{\partial \bar{u}_i}$ the operator $\sum_{\alpha=1}^{n} \frac{\partial}{\partial v_j^{\alpha}}\frac{\partial}{\partial \bar{u}_i^{\alpha}}$.

Under the map A the vector field $\frac{\partial}{\partial \xi_j}$ is transformed into $-i(\frac{1}{2}\bar{z}_j + \frac{\partial}{\partial \bar{z}_j})$ and the multiplication ξ_j is transformed into

$$-i(\frac{\partial}{\partial \bar{z}_j} - \tfrac{1}{2}\bar{z}_j) .$$

Hence we see that a function $\varphi \in \mathcal{H}_{p,q}$ is invariant under $O(p,q)$, if and only if

α) φ is invariant by the natural action of $O(p) \times O(q)$ on $M_{n,p,q}(\mathbb{C})$.

β) φ satisfies the equations $(\frac{\partial}{\partial v_j}\frac{\partial}{\partial \bar{u}_i} - \frac{1}{4} v_j\bar{u}_i)\cdot\varphi = 0$.

3.8 We define $\mathcal{O}(p,q)(0) \subset \mathcal{O}(p,q)$. $\mathcal{O}(p,q)(0)$ consists of the space of holomorphic functions in u , antiholomorphic in v , such that

$$\left(\frac{\partial}{\partial u_i}\frac{\partial}{\partial \bar{v}_j} - \frac{1}{4} u_i \bar{v}_j\right)\cdot \varphi = 0 \qquad \forall\, i,j \ .$$

As $O(p,q)$ acts unitarily on $\mathcal{H}_{p,q}$, it is easy to see the

3.9 Lemma: Under the map $A'_{p,q}$, $\mathfrak{J}'(0)$ is identified to a submodule of $\mathcal{O}(p,q)(0)$.

Let $\tau \in \hat{K}$ and $\mathcal{O}(p,q)(\tau)$ the isotypic component of type τ under the action

$$(g\cdot f)(u,v) \;=\; (\det g)^{-(p-q)/2} f(g^{-1}u, g^{-1}v) \ .$$

Let us write f as a series $f = \Sigma f_i$ where each f_i is homogeneous of total degree i with respect to $(u,\bar{v})$. As the action of $U(n)$ respects the degree, we see that each f_i is of type τ .

Let $f \in \mathcal{O}_{p,q}(0)(\tau)$. We have $\frac{\partial}{\partial u_i}\frac{\partial}{\partial \bar{v}_j} - \frac{1}{4} u_i \bar{v}_j \cdot f = 0$. Let f_{n_0} be the term of lowest degree in the expansion of f as a series. It is clear that f_{n_0} satisfies the equation

$$\frac{\partial}{\partial u_i}\frac{\partial}{\partial \bar{v}_j}\cdot f_{n_0} \;=\; 0 \qquad \forall i,j \ .$$

Let us consider the action of $GL(n;\mathbb{C})$ on the complex polynomials P on $M_{n,p,q}(\mathbb{C})$ by

$$(g\cdot P)(u,v) = P(g^{-1}u, {}^t gv) .$$

The operators $A_{ij} = \frac{\partial}{\partial u_i}\frac{\partial}{\partial v_j}$ generates the set of all constant coefficients differential operators invariant under this action of $GL(n;\mathbb{C})$.

Let us consider $\mathbb{D}_{p,q} = \{P; \Delta_{ij}P = 0\}$. We have studied in [4] the decomposition of $\mathbb{D}_{p,q}$ under the action of $GL(n;\mathbb{C}) \times GL(p,\mathbb{C}) \times GL(q;\mathbb{C})$, $((g,g_1,g_2)\cdot P)(u,v) = P(g^{-1}ug , {}^t gvg_2)$
We recall here the results;
Let index a representation τ of $GL(n;\mathbb{C})$ by its highest weight

$$(n_1,n_2,\dots,n_j,0,\dots,0,-m_i,-m_2,\dots,-m_1)$$
$$n_j > 0 \ , \ m_i > 0 \ .$$

Then $\mathbb{D}_{p,q}(\tau) \neq 0$ if and only if $i \leq p$, $j \leq q$. In this case $\mathbb{D}_{p,q}(\tau)$ is irreducible under $GL(n,\mathbb{C}) \times GL(p,\mathbb{C}) \times (GL(q,\mathbb{C})$.
Its highest weight is

$$\underbrace{(n_1,n_2,\dots,n_j,0,\dots,0,-m_i,\dots,-m_2,-m_1)}_{n} \otimes \underbrace{(m_1,m_2,\dots,m_i,0\dots0)}_{p}$$

$$\otimes \underbrace{(n_1,n_2,\dots,n_j,0,\dots,0)}_{q} \ .$$

Let $\lambda \in K^{\wedge}$ indexed by its highest weight. We introduce $C_{p,q} \subset K^{\wedge}$ given by

$$C_{p,q} = \{\lambda = (m_1, m_2, \ldots, m_i, 0 \ldots 0, -n_j, \ldots, -n_1)\ ,\ i \leq p,\ j \leq q\} .$$

We define $\delta_{p,q}$ the one dimensional representation of K given by $g \to (\det g)^{\frac{1}{2}(p-q)}$.

We can conclude from the preceeding remarks:

3.10: If $\mathfrak{J}(0)(\tau') \neq 0$ then $\tau' \in \delta_{p,q} + C_{p,q}$.

Hence we obtain the following results:

3.11 <u>Proposition</u>. If $H_{p,q}(\tau) \neq 0$, then $\tau \in \delta_{p,q} + C_{p,q}$.

Let us remark that for $p \geq n$ and $q \geq n$, then $C_{p,q} = \hat{K}$. Otherwise $C_{p,q}$ is a proper subset of $\hat{K}$.
We introduce the partition of $\hat{K}$ given by the number of positive signes of its highest weight and the number of negative signs.
I.e. for $p' + q' \leq n$ we define

$$\hat{K}(p',q') = \{(m_1, m_2, \ldots, m_{p'}, 0, \ldots, 0, -n_{q'}, \ldots, -n_1)\} \text{ with } m_i > 0$$
$$\text{with } m_i > 0\ ,\ n_j > 0\ .$$

We see that the K-support of the module $H(p,q)$ is incuded in $\delta_{p,q} + \bigcup_{\substack{p' \leq p \\ q' \leq q \\ p'+q' \leq n}} \hat{K}(p',q')$.

Comparing with 2.9, this suggest an analogy between the K-support of the module $H_{p,q}$ and the support of the Fourier transform.

In the article [7], we will give a relation between the asymptotic behavior of the K-types and the support of the Fourier transform, who will imply "asymptotically" the relation given here.

II. The group $U(n,n)$.

1. The domain G/K and its Shilov boundary.

We consider the complex vector space $\mathbb{C}^n \oplus \mathbb{C}^n = (V_1 \oplus V_2)$ with basis $e_1,e_2,\ldots,e_n,f_1,f_2,\ldots,f_n$. We write any $2n\times 2n$ complex matrix x by blocs $x = \left(\begin{array}{c|c} x_{11} & x_{12} \\ \hline x_{21} & x_{22}\end{array}\right)$.

Let us consider the hermitian matrix $h = \left(\begin{array}{c|c} 0 & -i \\ \hline i & 0\end{array}\right)$. We denote by (x,y) the canonical complex symmetric bilinear form on $\mathbb{C}^{2n}$, and by $\langle x,y\rangle = (x,\bar{y})$ the canonical hermitian form. Then $h(x,y) = \langle hx,y\rangle$ is an hermitian form on $\mathbb{C}^n \oplus \mathbb{C}^n$ of signature (n,n) . We consider the group $G = U(n,n) = \{g \in GL(2n;C);\ g^*hg = h\}$.

We consider X the complex grassmannian of n-dimensional subspace λ of $\mathbb{C}^n \oplus \mathbb{C}^n$. We consider G as imbedded in $GL(2n;\mathbb{C})$ which acts homogeneously on X in the obvious way. Let D be the open subset of X consisting of all λ's such that the hermitian form h is negative definite on λ . Let $\lambda \in D$, then

$$\lambda \cap (\mathbb{C}e_1 + \mathbb{C}e_2 + \ldots + \mathbb{C}e_n) = \{0\} ,$$

i.e. $$\lambda = \{w + zw \ ;\ w \in V_2 \ ;\ z\colon V_1 \to V_2\} .$$

Let us write $z = x+iy$ with $x = x^*$, $y = y^*$. The condition

$h(w+zw,w+zw) < 0$ is equivalent to $y > 0$.

We will identify the complex manifold D with the open subset of all the $(n\times n)$ - complex matrices defined by $D = \{z = x+iy;\ x = x^*,\ y = y^*,\ y > 0\}$. The group G acts homogeneous on D by $g\cdot z = (az+b)(cz+d)^{-1}$ if $g = \left(\begin{array}{c|c} a & b \\ \hline c & d \end{array}\right)$.

Let us consider $H(n)$ the vector space of all $n\times n$ hermitian matrices $x = x^*$. We can consider $H(n)$ as the Shilov boundary of D via $H(n) = \overline{D} \cap (y=0)$. The action of G on D gives rise to an action of G on $H(n)$ defined almost everywhere by $g\cdot x = (ax+b)(cx+d)^{-1}$. The Jacobian of this transformation is $\det(cx+d)^{-2n}$. We consider the unitary representation T of G in $L^2(H(n))$ defined by

$$(T(g)f)(x) = (\det(cx+d))^{-n} f((ax+b)(cx+d)^{-1})$$

$$\text{for } g^{-1} = \left(\begin{array}{c|c} a & b \\ \hline c & d \end{array}\right) .$$

Let us consider the Parabolic subgroup $P = \left(\begin{array}{c|c} a & * \\ \hline 0 & (a^*)^{-1} \end{array}\right)$ of $U(n,n)$. Each element of P can be written uniquely $P = \left(\begin{array}{c|c} a & 0 \\ \hline 0 & (a^*)^{-1} \end{array}\right)\left(\begin{array}{c|c} 1 & x \\ \hline 0 & 1 \end{array}\right)$ with $a \in GL(n,\mathbb{C})$ and $x \in H(n)$.

We first decompose the unitary representation T with respect to P . Let us consider the bilinear form $\operatorname{Tr} xy$ on $H(n)$ and the Fourier transform

$$f(y) = \int e^{-i \operatorname{Tr} xy} f(x)\, dx .$$

Then the action of P is given, after Fourier transform by

$$(T\begin{pmatrix} a & 0 \\ 0 & (a^*)^{-1} \end{pmatrix}\hat{f})(y) = (\det a)^n f(a\, y\, {}^t a)$$

$$(T\begin{pmatrix} 1 & x \\ 0 & 1 \end{pmatrix}\hat{f})(y) = e^{i \operatorname{Tr} xy} f(y) .$$

Let us consider the action of $GL(n,\mathbb{C})$ on $H(n)$ given by $a\cdot y = aya^*$. Then $H(n)$ is a finite union of orbits under $GL(n,\mathbb{C})$ given by the signature of the matrix y. We have consider for $k_1 + k_2 \leq n$, the set $\mathcal{O}_{k_1,k_2}$ of hermitian matrices of signature (k_1,k_2). For $k_1 + k_2 = n$, $\mathcal{O}_{k_1,k_2}$ is an open set of $H(n)$. It is clear from the preceding formulas that for $k_1 + k_2 = n$ the subspace $L^2_{k_1,k_2}$ of functions $f \in L^2(dx)$ whose Fourier transform is supported in $\overline{\mathcal{O}_{k_1,k_2}}$ is an irreducible invariant subspace of the representation $T|P$.

We denote by T_{k_1,k_2} the restriction of $T|P$ to the subspace $L^2_{k_1,k_2}$. Hence we have

1.] $T|P = \bigoplus_{k_1+k_2=n} T_{k_1,k_2}$ as a multiplicity free direct sum of irreducible representations of P.

2.- <u>The harmonic representation of</u> $U(n,n)$.

We define the following elements of $U(n,n)$:

$$g(a) = \begin{pmatrix} a & 0 \\ 0 & (a^*)^{-1} \end{pmatrix} \quad \text{for } a \in GL(n,C) ,$$

$$t(b) = \begin{pmatrix} 1 & b \\ 0 & 1 \end{pmatrix} \quad \text{for } b \in H(n) ,$$

$$\sigma = \begin{pmatrix} 0 & 1 \\ -1 & 0 \end{pmatrix}$$

These elements generate $U(n,n)$.

The following choice of $L(g)$ determines a representation of $U(n,n)$ in $L^2(\mathbb{C}^n)$:

$$(L(g(a))f)(w) = (\det a)f(a^*w)$$

$$(L(t(b))f)(w) = e^{-i\langle bw,w\rangle}f(w)$$

$$(L(\sigma)f)(w) = (\frac{i}{2\pi})^n \int e^{2i \operatorname{Re}\langle w,w'\rangle} f(w') \, |dw'|^2 ,$$

i.e. $L(\sigma)$ is the Fourier transform on $L^2(\mathbb{C}^n)$.

We consider the representation $\bar{L}$ of G in $L^2(\mathbb{C}^n)$ and the representation $L_{p,q} = \overset{p}{\otimes} L \otimes \overset{q}{\otimes} \bar{L}$ of G on $L^2(\mathbb{C}^{n\times(p+q)})$. Let $M_{n,p,q}(\mathbb{C}) = \operatorname{Hom}_{\mathbb{C}}(\mathbb{C}^{p+q},\mathbb{C}^n)$. We write an element $\xi: \mathbb{C}^{p+q} \to \mathbb{C}^n$ as $\xi = (\alpha, \beta)$ when $\alpha: \mathbb{C}^p \to \mathbb{C}^n$, $\beta: C^q \to C^n$.

We consider the canonical hermitian form Q on $\mathbb{C}^{p+q}$ of signature (p,q) , the $\xi Q \xi^*$ is a $n\times n$ hermitian matrix which is equal to $\alpha\alpha^* - \beta\beta^*$.

$L_{p,q}$ is realized in $L^2(M_{n,p,q})$ as follows:

$$(L_{p,q}(g(a))f)(\xi) = (\det a)^p(\det a^*)^q f({}^t a\xi)$$

$$(L_{p,q}(t(b))f)(\xi) = e^{-i\,\mathrm{Tr}\,(\xi Q\xi^* b)} f(\xi)$$

$$(L_{p,q}(\sigma)f)(\xi) = \left(\frac{i}{2\pi}\right)^{(p-q)n} \int e^{2i\,\mathrm{Re}\,\mathrm{Tr}(\xi Q\xi'^*)} f(\xi')|d\xi'|^2 .$$

On $L^2(M_{n,p,q})$ the action of the group $U(p,q)$ given by $(h\cdot f)(\xi) = f(\xi h)$ commutes with the representation $L_{p,q}$ of G.

We recall [4] that from $q = 0$ the operator

$$(\mathcal{I}_{p,o}\varphi)(x) = \int e^{i\,\mathrm{Tr}\,(\xi\xi^*)x}\, \varphi(\xi)d\xi$$

intertwines the representation $L_{p,o}$ with the representation

$$(T_p(g)f)(x) = \det(cx+d)^{-p}\, f((ax+b)(cx+d)^{-1})$$

i.e. for $\varphi \in \mathcal{S}(M_{n,p})$, $\mathcal{I}_{p,o}\varphi$ is a C^∞ vector of the representation T_p, and this map commutes with the action of G.

We recall that for $g \in SU(n,n)$ the functions $\overline{\det(cx+d)}$ is real. Hence for $g \in SU(n,n)$ we have $\det(cx+d)^p\overline{\det(cx+d)}^q = (\det(cx+d))^{p+q}$.

We consider as in (I.2) the operator $\mathcal{I}_{p,q}$

$$(\mathcal{I}_{p,q}\varphi)(x) = \int_{M_{n,p,q}} e^{i\,\mathrm{Tr}\,\xi Q\xi^* x}\, \varphi(\xi)d\xi .$$

For simplicity, we will consider here only the case $p+q = n$.

As in I.2, we obtain

2.1 Proposition. Let $p+q = n$, then $\mathcal{F}_{p,q}$ sends $\mathfrak{J}(M_{n,p,q})$ in $L^2(H(n),dx)$ and intertwines the representation $L_{p,q}$ with the representation T .

Now let us consider $L^2(H(n)) = \bigoplus_{p+q=n} L^2_{p,q}$, where $L^2_{p,q}$ denotes the space of functions f in $L^2(dx)$ whose Fourier transform are supported on the set of matrices of signature (p,q) .

From the proposition 2.1, we conclude that $L^2_{p,q}$ is a G-invariant subspace of $L^2(H(n))$ with respect to the representation T . Using 1.1, we conclude here directly the

2.2 Theorem. The representation T of G in $L^2(H(n))$ is the direct sum of the distinct irreducible representations $T_{p,q}$ of G in $L^2_{p,q}$.

References

[1] R. Howe: On some results of Strichartz and of Rallis and Schiffmann. To appear in Journal of Functional Analysis.

[2] H. Jakobsen and M. Vergne: "The Wave and Dirac operators and representations of the conformal group". J. Funct. Anal. vol 24 (1977), pp. 52-106.

[3] Kashiwara, M., Kimura, T. and Muro, M.: Microlocal calculus of simple holonomic system and its applications. (To appear.) See also Kashiwara, M. and Miwa, T.: Microlocal calculus and Fourier transformation of relative invariants of prehomogeneous vector space: Surikaiseki kenkyujo kokyuroku, 238 (1975), pp. 60-147 (in Japanese).

[4] M. Kashiwara and M. Vergne: On the Segal-Shale-Weil Representations and Harmonic Polynomials.. Inventiones math. 44 (1978), pp. 1-47.

[5] S. Rallis and G. Schiffmann: Discrete spectrum of the Weil Representation. Bull. Amer. Math. Soc. 83 (1977), pp. 267-270.

[6] H. Rossi and M. Vergne: Equations de Cauchy-Riemann tangentielles associées à un domaine de Siegel. Annales Scientifiques de l'Ecole Normale Superieure, vol. 9, (1976), pp. 31-80.

[7] M. Kashiwara and M. Vergne: K-types and singular spectrum. (same volume).

K-types and Singular Spectrum

by

M. Kashiwara

Research Institute for Mathematical Sciences
University of Kyoto

and

M. Vergne*

*Centre National de la Recherche Scientifique
Massachusetts Institute of Technology, MCS 78-02969

Introduction.

In "Functions on the Shilov boundary of the generalized half plane" in the same volume, we constructed unitary representations T_{k_1,k_2} of the Lie group $Sp(n,\mathbb{R})$. T_{k_1,k_2} is realized on the space of functions $u(x)$ on the space $S(n)$ of $n \times n$ symmetric matrices such that the Fourier transform

$$\hat{u}(\xi) = \int_{S(n)} u(x)\, e^{-2i\pi \operatorname{Tr} \xi x}\, dx$$

is supported on the set of symmetric matrices of signature (k_1,k_2) . On the same time, we observed that the K-types of T_{k_1,k_2} are distributed on a cone closely connected with (k_1,k_2) .

Why there is a relation between K-types and the Fourier transform?

More generally, let us consider G a real semi-simple Lie group, K its maximal compact subgroup, P a parabolic of G such that G/P is a symmetric space. Then $P = MN$ where N is abelian. We can then write, up to a set of measure zero, $G = N^-P$ where N^- is the nilpotent subgroup opposite to N . We hence can consider the compact manifold $G/P = X$ as an homogeneous space under K , or as an "almost" homogeneous space under $\mathfrak{n}^-$, the Lie algebra of N^- . Hence we can analyze $L^2(K/M\cap P) = L^2(X)$ either by the Fourier integral $\hat{u}(\xi)$ on $\mathfrak{n}$

$$\hat{u}(\xi) = \int_{\mathfrak{n}^-} u(x) e^{-2i\pi\langle\xi,x\rangle}\, dx$$

(we identify $(\mathfrak{n}^-)^*$ with $\mathfrak{n}$ by the killing form), or via the Fourier series expansion $u(k) = \Sigma\, u_\lambda(k)$ of u with respect to the finite dimensional representations of K .

Let us consider the singular spectrum SSu of the generalized function u . This subset of T^*X indicates the directions where u can be continued as an holomorphic functions on the complexification $X_{\mathbb{C}}$ of X .

If the Fourier transform $\hat{u}(\xi)$ is supported on a closed cone Γ , then the singular spectrum of $u(x)$ is contained in the subset $N_- \times \sqrt{-1}\,\Gamma$ of $\sqrt{-1}$ times the cotangent bundle $N_- \times \eta = T^*N_-$.

In this note, we prove a similar relation between the K-types appearing in the expansion of u and the singular spectrum of u .

The determination by Birgit Speh of the K-types of solutions of mass zero equations on the Minkowski-space considered as an homogeneous space under $U(2,2)$ was our first indication that there was a strong connection between the asymptotic behavior of the K-types of a given representation T of G and the geometric realization of T as acting on functions on the Minkowski space solutions of differential equations.

These questions are also in strong relation with the orbit method: Let G be a semi-simple Lie group and K a compact subgroup of G . Let $\mathcal{O}$ an orbit of G in $\mathfrak{g}^*$ and $\rho_{\mathcal{O}}$ the representation of G which can be in numerous cases associated to $\mathcal{O}$. Then the asymptotic directions to the K-types occurring in $\rho_{\mathcal{O}}$ should be the projection of the asymptotic cone of the orbit $\mathcal{O}$. In particular we prove that the asymptotic support of the K-types of an arbitrary Harish-Chandra module is given by the projection on k^* of nilpotent orbits of G in $\mathfrak{g}^*$, hence are only among a finite number of possibilities.

We thank Dan Barbasch for several discussions on asymptotic directions of the orbits of G in $\mathfrak{g}^*$ and Sigurdur Helgason for discussions on asymptotic estimates of the spherical functions.

1. Let K be a connected compact Lie group and H a Cartan subgroup of K. Let $\mathfrak{k}$ and $\mathfrak{h}$ be the Lie algebras of K and H. We denote by $\mathfrak{k}_{\mathbb{C}}$ and $\mathfrak{h}_{\mathbb{C}}$ their complexifications. We fix a K-invariant metric on $\mathfrak{k}$, by which we identify $\mathfrak{k}$ and $\mathfrak{h}$ with their dual vector spaces. Let $(\ ,\ )$ be the hermitian metric on $\mathfrak{k}_{\mathbb{C}}$ induced by this metric on $\mathfrak{k}$.

For any $\alpha \in \sqrt{-1}\,\mathfrak{h}^*$, we define

$$\mathfrak{k}_\alpha = \{X \in \mathfrak{k}_{\mathbb{C}} \,;\, [H,X] = \alpha(H)X \text{ , for } H \in \mathfrak{h}\} .$$

Then $\dim \mathfrak{k}_\alpha = 0$ or 1, for $\alpha \neq 0$, and $\mathfrak{k}_\alpha = \mathfrak{h}_{\mathbb{C}}$ for $\alpha = 0$.

Let us denote $\Delta = \{\alpha \in \sqrt{-1}\,\mathfrak{h}^* - \{0\}$ such that $\mathfrak{k}_\alpha \neq 0\}$ the set of roots, and let Δ^+ be the set of positive roots with respect to some ordering. We define $\mathfrak{n} = \bigoplus_{\alpha\in\Delta^+} \mathfrak{k}_\alpha$. Then $\mathfrak{h}_{\mathbb{C}} + \mathfrak{n}$ is a Borel subalgebra of $\mathfrak{k}_{\mathbb{C}}$.

We have

(1.1) $\mathfrak{k}_{\mathbb{C}} = \mathfrak{k} \oplus \sqrt{-1}\,\mathfrak{h} \oplus \mathfrak{n}$, considered as a sum of real vector spaces.

Let L be the lattice of $\sqrt{-1}\,\mathfrak{h}^*$ which comes from the character group of H, i.e. the set of $\alpha \in \sqrt{-1}\,\mathfrak{h}^*$ such that $\chi(e^H) = e^{\alpha(H)}$ $(H \in \mathfrak{h})$ defines a character on H. Let C be the Weyl chamber:

$$C = \{\lambda \in \sqrt{-1}\,\mathfrak{h}^* \,;\, \langle\lambda,\alpha\rangle > 0 \ , \ \forall\, \alpha \in \Delta^+\} .$$

Then the set $\hat{K}$ of irreducible representations of K is isomorphic to $L \cap \overline{C}$ by the highest weight. For $\lambda \in L \cap \overline{C}$, let V_λ denote the irreducible representation with highest weight λ.

2. Let dk be the Haar measure on K normalized by $\int_K dk = 1$.

Then $L^2(K)$, considered as a $K \times K$ module by left and right translations, has the following decomposition into irreducible components:

(2.1) $$L^2(K) = \bigoplus_{\lambda \in L \cap \overline{C}} (V_\lambda \otimes V_\lambda^*) ,$$ the element $v \otimes f$ of $V_\lambda \otimes V_\lambda^*$ being identified with the real analytic function $k \longrightarrow f(k^{-1} \cdot v)$ on K

For φ_λ an element of $V_\lambda \otimes V_\lambda^*$, we denote by $\|\varphi_\lambda\|$ the norm induced by $L^2(K)$.

2.2 Theorem [1], [10]:

Let φ be a function on K and develop $\varphi = \sum_{\lambda \in L \cap \overline{C}} \varphi_\lambda$, then

1) $\varphi = \Sigma \varphi_\lambda$ is a real analytic function on K if and only if there are positive constants C and δ such that

$$\|\varphi_\lambda\| \leq Ce^{-\delta|\lambda|} .$$

2) $\varphi = \Sigma \varphi_\lambda$ is a C^∞-function, if and only if, for any positive integer m , there exists a positive number C_m such that

$$\|\varphi_\lambda\| \leq C_m(1 + |\lambda|)^{-m} .$$

3) $\varphi = \Sigma \varphi_\lambda$ is contained in $L^2(K)$ if and only if

$$\Sigma \|\varphi_\lambda\|^2 < \infty .$$

4) $\varphi = \Sigma \varphi_\lambda$ is a distribution if and only if there are positive numbers m and C such that

$$\|\varphi_\lambda\| \leq C(1 + |\lambda|)^m$$

5) $\varphi = \Sigma \varphi_\lambda$ is a hyperfunction if and only if for any $\varepsilon > 0$, there exist $C_\varepsilon > 0$ such that

$$\|\varphi_\lambda\| \leq C_\varepsilon \; e^{\varepsilon|\lambda|}$$

(For the theory of microfunctions, we refer to [2], [3], [4].)

3. Let TK and T^*K be the tangent and cotangent vector bundles of K. We shall identify TK with $K \times \mathfrak{k}$ and T^*K with $K \times \mathfrak{k}^*$ by left translations. Therefore, the right (resp. left) translation of $k_0 \in K$ on K give rise to the transformation on $TK = K \times \mathfrak{k}$: $(k,X) \longrightarrow (kk_0,X)$ (resp: $(k,X) \longrightarrow (k_0k \, , \, \mathrm{Ad}\, k_0 \cdot X)$) and on $T^*K = K \times \mathfrak{k}^*$: $(k,\xi) \longrightarrow (kk_0,\xi)$ (resp: $(k,\xi) \longrightarrow (k_0k \, , \, \mathrm{Ad}^*\, k_0 \cdot \xi)$). Here Ad^* is the coadjoint representation.

Let T be a closed cone of $\sqrt{-1}\,\mathfrak{h}^*$ contained in $\overline{C}$ and let $\varphi = \sum_{\lambda \in T \cap L} \varphi_\lambda$ be a hyperfunction on K such that each φ_λ is a highest weight vector of highest weight λ with respect to the left action on $L^2(K)$. Then we have

$$X \cdot \varphi_\lambda = 0 \qquad \text{for } x \in \mathfrak{n}$$

$$X \cdot \varphi_\lambda = \lambda(X)\varphi_\lambda \qquad \text{for } x \in \mathfrak{h}$$

for the left action.

Let $K_{\mathbb{C}}$ be a complexification of K, then for some $\rho > 0$ the map f_ρ defined on

$$U_\rho = \{(k,H,X) \in K \times \mathfrak{h} \times \mathfrak{n} \, ; \, |H| < \rho \, , \, |X| < \rho\}$$

by $f_\rho(k,H,X) = \exp(\sqrt{-1}\, H + X)\, k$ is an isomorphism from U_ρ onto an open neighborhood Ω_ρ of K in $K_{\mathbb{C}}$ as follows from (1.1).

Let us extend the function φ_λ holomorphically on $K_{\mathbb{C}}$. We fix H and X and we consider the function φ_λ on the translate $\exp(\sqrt{-1}\, H + X) \cdot K$ of K in $K_{\mathbb{C}}$. Then

$$\varphi_\lambda(\exp(\sqrt{-1}\,H + X)\cdot k) = e^{-\langle\lambda,\sqrt{-1}\,H\rangle}\,\varphi_\lambda(k)\ .$$

Hence considered as a function of k, we have:

$$\int |\varphi_\lambda(\exp(\sqrt{-1}\,H + X)\,k)|^2\,dk = e^{-2\langle\lambda,\sqrt{-1}\,H\rangle}\,\|\varphi_\lambda\|^2$$

We define $T^0 = \{H \in \mathfrak{h}\ ;\ \langle\lambda,\sqrt{-1}\,H\rangle > 0, \forall\lambda \in T \subset \sqrt{-1}\,\mathfrak{h}^*\}$ and $T^0_\rho = \{(k,H,X)\ ;\ H \in T^0\} \cap U_\rho$. Then $\Sigma\,\varphi_\lambda((\exp\sqrt{-1}\,H + X)\,k)$ converges on T^0_ρ: φ_λ being an hyperfunction satisfies $\|\varphi_\lambda\| \leq C_\varepsilon e^{\varepsilon|\lambda|}$ and hence $e^{-\langle\lambda,\sqrt{-1}\,H\rangle}\,\|\varphi_\lambda\|$ has exponential decay for $H \in T^0$. Hence

$$\varphi(\exp(\sqrt{-1}\,H + X)k) = \Sigma\,\varphi_\lambda(\exp(\sqrt{-1}\,H + X)k)$$

is a holomorphic function defined on $f_\rho(T^0_\rho)$. This domain is an infinitesimal neighborhood of $\{(k,\sqrt{-1}\,X \in \sqrt{-1}\,TK$, with $\sqrt{-1}\,X \in (\sqrt{-1}\,T^0 + \mathfrak{n} + \mathfrak{k}) \cap \sqrt{-1}\,\mathfrak{k}\}$. Let us consider $\mathfrak{h}^\perp$ the orthogonal complement of $\mathfrak{h}$ in $\mathfrak{k}$. We have $(\mathfrak{h}^\perp)_{\mathbb{C}} = \mathfrak{h}^\perp \oplus \mathfrak{n}$ as a sum of real vector subspaces. Hence

$$(\sqrt{-1}\,T^0 + \mathfrak{n} + \mathfrak{k}) \cap \sqrt{-1}\,\mathfrak{k} = \sqrt{-1}\,T^0 + \sqrt{-1}\,\mathfrak{h}^\perp\ .$$

Hence φ converges on an infinitesimal neighborhood of (k,X), for $X \in \sqrt{-1}\,T^0 + \sqrt{-1}\,\mathfrak{h}^\perp$. By definition, the singular spectrum $SS\,\varphi$ of φ is then contained in the dual of this neighborhood in $\sqrt{-1}\,T^*K$.

We imbed $\mathfrak{h}^*$ in $\mathfrak{k}^*$ according to the decomposition $\mathfrak{k}^* = \mathfrak{h}^* \oplus (\mathfrak{h}^\perp)^*$. Then it is immediate from the definition of T^0 that the dual cone of $T^0 + \mathfrak{h}^\perp$ in $\mathfrak{k}^*$ is the convex hull of $\sqrt{-1}\,T$. Thus we obtain

$$SS\,\varphi \subset \{(k,\xi)\ ;\ -\xi \in \text{convex hull of } T\}\ .$$

3.1 Proposition: Let T be a closed cone in $\sqrt{-1}\,\mathfrak{h}^* \cap \overline{C}$. Suppose that $\varphi = \sum_{\lambda \in T} \varphi_\lambda$ is an hyperfunction on K and all the φ_λ are highest weight vectors with respect to the left action. Then $SS\,\varphi \subsetneq -K\cdot T$, where $T \subset \sqrt{-1}\,\mathfrak{k}^* = \sqrt{-1}\,T_eK^*$ and K acts by the right action on T^*K.

Proof: If T is convex, this follows from the preceding discussion. In the general case, for any disjoint family $\{T_j\}_{1 \le j \le N}$ of closed convex sets such that $T \subset \cup T_j$, we have $\varphi = \Sigma\,\varphi_j$ with $\varphi_j = \sum_{\lambda \in T_j} \varphi_\lambda$ and $SS\,\varphi_j \subset -K\cdot T_j$. Therefore

$$SS\,\varphi \subset \cup(-K\cdot T_j)$$
$$= -K\cdot(\cup T_j)\ .$$

Since $\cup T_j$ can be as close to T as we like, we obtain the result.

4. Singular support and K-types.

Let χ_λ be the character of V_λ, i.e. $\chi_\lambda(k) = tr(\tau_\lambda(k); V_\lambda)$. We know that $\{\chi_\lambda\}$ forms an orthogonal basis of the space of $(Ad\ K)$-invariant L^2-functions on K. χ_λ is the unique $(Ad\ K)$-invariant functions in $V_\lambda^* \otimes V_\lambda$.

Let us consider the δ-function $\delta(k)$ supported at e characterized by $\int \delta(k)u(k)dk = u(e)$. We have $\delta(k) = \sum_\lambda \varphi_\lambda$, with $\varphi_\lambda \in V_\lambda \otimes V_\lambda^*$. Since $\delta(k)$ is invariant by $Ad\ K$, φ_λ is proportional to $\overline{\chi}_\lambda$. Since $(\varphi_\lambda, \overline{\chi}_\lambda) = \int \delta(k)\chi_\lambda(k)dk = \chi_\lambda(e) = \dim V_\lambda$, we obtain $\varphi_\lambda = \dim V_\lambda\,\overline{\chi}_\lambda$, i.e. $\delta = \sum_{\lambda \in L \cap \overline{C}} (\dim V_\lambda)\,\chi_\lambda$.

Let u_λ be the highest weight vector of the representation V_λ, provided with a K-invariant hermitian inner product $(\ ,\ )$. We set $\psi_\lambda(k) = (\tau_\lambda(k^{-1})u_\lambda, u_\lambda)$. We shall calculate $\int \psi_\lambda(k'kk'^{-1})dk'$. For any u and v $\int (\tau_\lambda(k'k^{-1}k'^{-1})u, v)dk'$ is an $Ad\ K$-invariant function contained in $V_\lambda \otimes V_\lambda^*$ and hence is proportinal to $\overline{\chi}_\lambda$.

Hence there exist a constant c such that

$$\int (\tau_\lambda(k'k^{-1}k'^{-1})u,v)\,dk' = c\,(u,v)\,\overline{\chi}_\lambda(k) .$$

Setting $k = e$, we have $c = \dfrac{1}{\overline{\chi}_\lambda(e)} = \dfrac{1}{\dim V_\lambda}$ i.e.

$$\int (\tau_\lambda(k'k^{-1}k'^{-1})u,v)\,dk' = \frac{1}{\dim V_\lambda}(u,v)\,\overline{\chi}_\lambda(k) .$$

If we normalize u_λ by $\|u_\lambda\| = 1$, we have

(4.1) $$\int_K \psi_\lambda(k'kk'^{-1})\,dk' = \frac{1}{\dim V_\lambda}\overline{\chi}_\lambda(k) .$$

We define for a cone T in $\overline{\mathcal{C}}$

$$\psi_T = \sum_{\lambda\in T\cap L} (\dim V_\lambda)^2\,\psi_\lambda$$

and

$$\delta_T = \sum_{\lambda\in T\cap L} (\dim V_\lambda)\,\overline{\chi}_\lambda .$$

By (2.2), ψ_T and δ_T are hyperfunctions. By (4.1), δ_T is obtained from ψ_T by

$$\delta_T(k) = \int \psi_T(k'kk'^{-1})\,dk' .$$

Since $SS\,\psi_T$ is contained in $-K\cdot T$ by proposition 3.1, we obtain

(4.2) $$SS\,\delta_T \subset -(K\times K)\cdot T .$$

<u>4.3. Lemma</u>: Any $K\times K$ invariant subset of $\sqrt{-1}\,T^*K$ is of the form $-(K\times K)\cdot T$ with T a subset of $\overline{\mathcal{C}}$.

<u>Proof</u>: This is equivalent to the classification of Ad K-invariant sets of $\mathcal{K}$, and it is well known that they are written in the form $\mathrm{Ad}\,K\cdot T$ with $T\subset\overline{\mathcal{C}}\subset\sqrt{-1}\,\mathfrak{h}^*$ for a unique T.

Let $u(k)$ and $v(k)$ be two hyperfunctions on K. We define their convolution $u * v$ by $(u * v)(k) = \int_{h \in K} u(kh^{-1})\, v(h)dh$.

We have
$$\begin{aligned} \dim V_\lambda\, \overline{\chi}_\lambda * u &= u \quad \text{for} \quad u \in V_\lambda \otimes V_\lambda^* \\ &= 0 \quad \text{for} \quad u \in V_{\lambda'} \otimes V_{\lambda'}^*, \\ & \qquad\qquad\quad \lambda \neq \lambda' . \end{aligned}$$

Hence for a hyperfunction $\varphi = \sum_{\lambda \in L \cap \overline{C}} \varphi_\lambda$, we have $\delta_T * \varphi = \sum_{\lambda \in L \cap T} \varphi_\lambda$

4.4 Lemma: Let T_1 and T_2 be two closed cones in $\overline{C}$. If $SSu \subset -(K \times K) \cdot T_1$, and if $SSv \subset -(K \times K) \cdot T_2$, then we have

$$SS(u * v) \subset -\ (K \times K)(T_1 \cap T_2) .$$

Proof: This lemma is easily derived from the behavior of the singular spectrum under integration: The singular spectrum of $u(kh^{-1})$ considered as a hyperfunction on $K \times K$ is contained in

$$\{(k,h;\xi\ ,\ -\ Ad^*(kh^{-1})^{-1}\cdot\xi)\ ;\ (kh^{-1};\xi) \in SSu\} .$$

Therefore the singular spectrum of $u(kh^{-1})v(h)$ is contained in

$$\{(k,h;\xi\ ,\ \xi' - Ad^*(kh^{-1})^{-1}\cdot\xi)\ ;\ \text{with}\ \ (kh^{-1};\xi) \in SSu\ ,\ (h,\xi') \in SSv\}$$

Hence the singular spectrum of $\int_K u(kh^{-1})\, v(h)dh$ is contained in

$$\{(k;\alpha)\ ,\ \text{such that there exists a}\ h\ \text{with}\ (k,h;\alpha,0) \in SS(u(kh^{-1})v(h))\} .$$

This implies $\alpha = \xi \in (Ad^*K)T_1$, $\xi' = Ad^*(kh^{-1})^{-1}\cdot\xi$. Hence $\alpha \in (Ad^*K)T_1 \cap (Ad^*K)T_2 = (Ad^*K)(T_1 \cap T_2)$.

Now, we are ready to prove the following theorem.

4.5 Theorem: Let $\varphi = \Sigma\, \varphi_\lambda$ be a hyperfunction on K . Let T be a closed cone in $\overline{C}$. Then the following conditions are equivalent:

(1) $SS\,\varphi \subset -(K \times K)\cdot T$.

(2) For any closed cone T' in $\overline{C}$ such that $T \cap T' \subset \{0\}$, there are constants $R_{T'} > 0$, and $\varepsilon_{T'} > 0$ such that

$$\|\varphi_\lambda\| \leq R_{T'} e^{-\varepsilon_{T'}|\lambda|} \quad \text{for } \lambda \in T' .$$

<u>Proof</u>: Let us prove first that (2) implies (1). Take T' as in (2), then $\varphi_{T'} = \delta_{T'} * \varphi = \sum_{\lambda \in T'} \varphi_\lambda$ and $\varphi_{\overline{C}-T'} = \delta_{\overline{C}-T'} * \varphi = \sum_{\lambda \notin T'} \varphi_\lambda$. By (2.2) and the hypothesis, $\varphi_{T'}$ is <u>real analytic</u>; by (4.2), (4.4), $SS\,\varphi_{\overline{C}-T'}$ is contained in $-K \times K\,(\overline{C} - T')$. Since we can take $\overline{C} - T'$ as close to T as we like, we obtain (1).

Reciprocally if (1) is satisfied, $SS\,\varphi_{T'} \subset -(K \times K)(T \cap T') = \{0\}$. Hence $\varphi_{T'}$ is a real analytic function. So (2) follows from Theorem (2.2).

<u>Remark</u>: If we employ the wave front set in the C^∞-sense instead of the singular spectrum in condition (1), then condition (2) must be changed to: For any $m > 0$, there is $C_m > 0$ such that

$$\|\varphi_\lambda\| \leq C_m (1 + |\lambda|)^{-m} \quad \text{for } \lambda \in T' .$$

5. K-types of induced representations.

Let M be a subgroup of K and $\mathfrak{m}$ the Lie algebra of M . Let X be the homogeneous space K/M . We denote by 0 the coset eM . Then the left action of K induces a surjective map: $\mathfrak{k} \longrightarrow T_0(X)$ whose kernel is $\mathfrak{m}$. Hence T_0^*X is identified with the orthogonal complement $\mathfrak{m}^\perp$ in $\mathfrak{k}^*$.

Let σ be a finite dimensional unitary representation of M in the complex vector space U . We denote by $\mathcal{U}$ the corresponding homogeneous vector bundle $K \times_M U$ over X . Hence the space of section of $\mathcal{U}$ is the space of U-valued functions $u(k)$ on K satisfying

$$(5.1) \qquad u(km) = \sigma(m)^{-1}u(k) \quad \text{for} \quad k \in K\,,\ m \in M\,.$$

The group K acts by left translations on this space. The space of L^2-sections of $\mathcal{U}$ is denoted by $L^2(K/M;\mathcal{U}) = L^2(X,U)$.

The decomposition of $L^2(X,U)$ under K is given by the Frobenius reciprocity law, i.e.:

$$(5.2) \qquad L^2(X,U) = \bigoplus_{\lambda \in \hat{K}} V_\lambda \otimes \mathrm{Hom}_M(V_\lambda, U)$$

where $v \otimes f$, for $v \in V_\lambda$, $f \in \mathrm{Hom}_M(V_\lambda, U)$ is identified with the function $(v \otimes f)(g) = f(g^{-1}v)$.

We denote by $W_\lambda = V_\lambda \otimes \mathrm{Hom}_M(V_\lambda, U)$.

We wish to determine what are the asymptotic behavior of the representations of K appearing in $L^2(X,U)$; i.e. what are the representations λ of K such that $\mathrm{Hom}_M(V_\lambda, U) \neq \{0\}$ when $|\lambda| \to \infty$. Consider the singular spectrum of a section u of $\hat{\mathcal{U}}$ regarded as a U-valued function on K satisfying 5.1. Since u(k) satisfies (5.1), we have:

$$SSu \subset \{(k,\xi);\ (Ad^*k^{-1})\xi \in \sqrt{-1}\ \mathfrak{m}^\perp\}\,.$$

We consider the inclusion $\mathfrak{m} \subset \mathfrak{k}$ and the corresponding map $p: \mathfrak{k}^* \to \mathfrak{m}^*$. The kernel of this map is $\mathfrak{m}^\perp$. We consider the set $(Ad^*K)\mathfrak{m}^\perp$ of orbits intersecting $\mathfrak{m}^\perp$. Let

$$(5.3) \qquad \mathfrak{h}^*_{\mathfrak{m}} = \mathfrak{h}^* \cap (Ad^*K)\mathfrak{m}^\perp\,.$$

Then every orbit intersecting $\mathfrak{m}^\perp$ intersects $\mathfrak{h}^*_{\mathfrak{m}}$

<u>5.4 Proposition</u>: For any closed cone T in $\overline{C}$ such that $T \cap \sqrt{-1}\,\mathfrak{h}^*_{\mathfrak{m}} \subset \{0\}$, there exists a constant R_T such that $\mathrm{Hom}_M(V_\lambda, V) = 0$ for $\lambda \in T$, and $|\lambda| \geq R_T$.

<u>Proof</u>: If it is not true, there is a sequence λ_j in T such that

$|\lambda_j|$ tends to infinity, when j tends to infinity, and such that $W_{\lambda_j} \neq \{0\}$. Let us take a vector φ_j in W_{λ_j} normalized by $\|\varphi_j\| = 1$. Take any sequence a_j in $\mathbb{C}$, such that $\Sigma |a_j|^2 < \infty$. We consider $u(k) = \Sigma a_j \varphi_j$ which belongs to $L^2(X,U)$. We have as u satisfies (5.1) $(SSu) \subset K\cdot(\sqrt{-1}\, \mathfrak{m}^{\perp})$ for the left action of K,

$$\subset (K \times K)(\sqrt{-1}\, \mathfrak{h}_{\mathfrak{m}}^*).$$

But by Theorem (4.5) as $T \cap \sqrt{-1}\, \mathfrak{h}_{\mathfrak{m}}^* = 0$, this would imply that there exist $R > 0$, and $\varepsilon > 0$ with $|a_j| \leq Re^{-\varepsilon|\lambda_j|}$. This cannot be true for any sequence a_j, with $\Sigma |a_j|^2 < \infty$, hence we obtain our result.

Remark: Let us consider $\hat{K}$ as a subset of orbits in $\sqrt{-1}\, \mathfrak{k}^*$ by $V_\lambda \longmapsto (Ad^*K)\cdot\lambda$. This is a bijection with the set of integral orbits of K in $\sqrt{-1}\, \mathfrak{k}^*$. Let us consider the projection of the orbit $\mathcal{O}_\lambda = (Ad^*K)\cdot\lambda$ on $\sqrt{-1}\, \mathfrak{m}^*$ with respect to the restriction $p: \sqrt{-1}\, \mathfrak{k}^* \to \sqrt{-1}\, \mathfrak{m}^*$. This set decomposes under M into a union of M-orbits. The "philosophy" of the orbit method would imply that the restriction of V_λ to M decomposes as a sum of representations μ_j of M corresponding to "some" integral orbits of M in $\sqrt{-1}\, \mathfrak{m}^*$ contained in the projection of $\mathcal{O}_\lambda$ on $\sqrt{-1}\, \mathfrak{m}^*$. In particular the λ's of $\hat{K}$ containing a given representation of M corresponds to orbits $\mathcal{O}_\lambda$ intersecting $p^{-1}(B)$ for B a compact subset of $\mathfrak{m}^*$. The asymptotic directions of the corresponding highest weights is $\sqrt{-1}\, \mathfrak{h}_{\mathfrak{m}}^* \cap \overline{C}$. Hence for a cone T such that $T \cap \mathfrak{h}_{\mathfrak{m}}^* = 0$, the set $(Ad^*K)\cdot T \cap p^{-1}(B)$ is a bounded set. Our result gives an "asymptotic" verification of this desired result. (We thank Donald King for discussions of the case $K \to K \times K$ via the diagonal map, i.e. of the case of decomposition of tensor products [5].)

We can reformulate our Theorem 4.5 in the following:

5.5 Theorem: Let $\varphi = \Sigma \varphi_\lambda$, $\varphi_\lambda \in W_\lambda$ be a hyperfunction section of $\mathcal{U}$. Let T be a closed cone in $\sqrt{-1}\, \mathfrak{h}_{\mathfrak{m}}^* \cap \overline{C}$. The following conditions are equivalent:

(1) $SS\varphi \subset -K\cdot(Ad^*K\cdot T \cap \sqrt{-1}\,\mathfrak{m}^{\perp})$ when K acts by the left.

(2) For any closed cone T' in $\sqrt{-1}\,\mathfrak{h}^*$ such that $T' \cap T = \{0\}$, there exists $R_{T'}$ and $\varepsilon_{T'}$ such that $\|\varphi_\lambda\| \leq R_{T'}e^{-\varepsilon_{T'}|\lambda|}$ for $\lambda \in T'$.

Remark: It is only necessary to investigate the condition (2) for the cones T' intersecting $\sqrt{-1}\,\mathfrak{h}^*_{\mathfrak{m}}$, as follows from 5.3.

The conditions of the Theorem (5.5) will be more easily described when K/M is a symmetric space. Let $\mathfrak{k} = \mathfrak{m} \oplus \mathfrak{p}$ the decomposition of $\mathfrak{k}$ with respect to the involution θ, i.e. $\theta|\mathfrak{m} = Id$, $\theta|\mathfrak{p} = -Id$. Let $\mathfrak{a}$ be a maximal abelian subalgebra of $\mathfrak{k}$ contained in $\mathfrak{p}$. We can choose a θ-stable Cartan subalgebra $\mathfrak{h}$ of $\mathfrak{k}$, such that $\mathfrak{h} \cap \mathfrak{p} = \mathfrak{a}$, i.e. $\mathfrak{h} = \mathfrak{h} \cap \mathfrak{m} \oplus \mathfrak{a}$. Let $C_{\mathfrak{a}}$ be a Weyl chamber of $\sqrt{-1}\,\mathfrak{a}^*$ and C a Weyl chamber of $\sqrt{-1}\,\mathfrak{h}^*$ compatible with $C_{\mathfrak{a}}$.

We define for $\mu \in \overline{C}_{\mathfrak{a}}$

$$F_\mu = \bigoplus_{\lambda|\mathfrak{a}=\mu} V_\lambda \otimes Hom_M(V_\lambda, U) .$$

We recall that if $Hom_M(V_\lambda, U) \neq 0$ then $\lambda|\mathfrak{h} \cap \mathfrak{m}$ is a weight of the representation U of M restricted to $\mathfrak{h} \cap \mathfrak{m}$ as follows from the following remark: Let f be a nonzero element of $Hom_M(V_\lambda, U)$ and v be the highest weight vector of V_λ. Clearly $f(v) \in U$ transform under $\mathfrak{h} \cap \mathfrak{m}$ according to $\lambda|\mathfrak{h} \cap \mathfrak{m}$. Hence we have to see that $f(v) \neq 0$. Let $\mathfrak{n}' = \{\bigoplus_{\alpha\in\Delta^+} \mathfrak{k}_\alpha, \alpha|\mathfrak{a} \neq 0\}$, we have $\mathfrak{k}^C = \mathfrak{m}^C + \mathfrak{a}^C + \mathfrak{n}'$. As v is an eigenvector for $\mathfrak{a}^C + \mathfrak{n}$, for any $u \in \mathcal{U}(\mathfrak{k}^C)$, we have $u\cdot v = u_0\cdot v$ with $u_0 \in \mathcal{U}(\mathfrak{m}^C)$. Hence $f(u\cdot v) = u_0\cdot f(v)$. As $\mathcal{U}(\mathfrak{k}^C).v = V_\lambda$, $f(v) \neq 0$.

In particular, for any μ, the possible λ's occuring in F_μ are of the form $\mu + \delta_j$ for a finite choice of δ_j in $\sqrt{-1}\,(\mathfrak{h} \cap \mathfrak{m})^*$. In this case we see that the possible K-types occuring in $L^2(K,U)$ are contained in a strip along $\mathfrak{a}^*$.

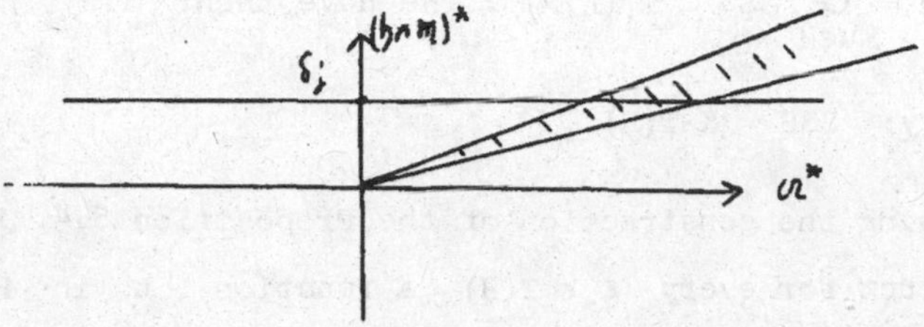

Hence the Proposition (5.4) is then automatically satisfied. We remark also that F_μ is finite dimensional.

Our Theorem (5.4) is reformulated as follows:

5.8 Theorem: Let $\varphi = \sum_{\mu \in \overline{C}_{\mathfrak{a}}} \varphi_\mu$ a hyperfunction section of $\mathcal{U}$ and S be a closed cone in $\overline{C}_{\mathfrak{a}}$. Then the following conditions are equivalent:

(1) $(SS\varphi) \subset -K \cdot S$

(2) For any closed cone S' contained in $\sqrt{-1}\,\mathfrak{a}^*$ satisfying $S \cap S' = \{0\}$, there are positive numbers $R_{S'}$ and $\varepsilon_{S'}$ such that $\|\varphi_\mu\| \leq R_{S'}\, e^{-\varepsilon_{S'}|\mu|}$ when $\mu \in S'$.

Proof: This follows from (5.5) as for a symmetric space, $(\mathrm{Ad}^*K)\mathfrak{m}^\perp \cap \overline{C} = \overline{C}_{\mathfrak{a}}$ and for any $S \subset \overline{C}_{\mathfrak{a}}$, $(\mathrm{Ad}^*K)S \cap \sqrt{-1}\,\mathfrak{m}^\perp = (\mathrm{Ad}^*M)S$.

Let $X = K/M$ being a symmetric space. Let us now consider H a K-invariant subspace of $L^2(X,U)$. We consider $H = \bigoplus_{\mu \in \overline{C}_{\mathfrak{a}}} H_\mu$ where

$$H_\mu = \bigoplus_{\lambda \in \hat{K}} H_\lambda \qquad \lambda|\mathfrak{a} = \mu .$$

5.9. We define the asymptotic support of H by

$T(H) = \{u \in \overline{C}_{\mathfrak{a}}$, such that there exist a sequence (u_n, ε_n) , $u_n \in \overline{C}_{\mathfrak{a}}$, $\varepsilon_n > 0$ with $|u_n| \to \infty$, $H_{u_n} \neq 0$ and $\varepsilon_n u_n \to u$.$\}$

We define $SSH = \bigcup_{u \in H} SSu \subset T^*(K/M)$. We have then:

5.10 Corollary: $SSH = K \cdot T(H)$.

Proof: Following the construction of the Proposition 5.4, it is easy to construct for every $\xi \in T(H)$ a function u in H such that $k \cdot (1, \xi) \in SSu$. As H is K-invariant the corollary follows.

6. Singular spectrum of G-modules.

Let G be a real semi-simple Lie group, K a maximal compact subgroup of G , $G = KAN$ an Iwasawa decomposition of G . We denote by M the centralizer of A in K . Let $(\mathfrak{g}, \mathfrak{k}, \mathfrak{a}, \mathfrak{n}, \mathfrak{m})$ be the Lie algebras of the groups involved.

For σ a finite dimensional representation of M in U and λ a homomorphism of A in C^* , we consider the representation $\sigma \otimes \lambda$ of MAN in the vector space U trivial on N and extending $\sigma \otimes \lambda$ on $M \times A$. We consider the G-bundle $G \times_{MAN} U$. This as a K-bundle is isomorphic to $K \times_M U$ over $X = K/M$. The decomposition under K of the associated principal series $\mathrm{Ind}_{MAN}^{G} \sigma \otimes \lambda$ is then given by the formula 5.2. The Proposition (5.4) gives us the asymptotic behavior of the K-types of the principal series associated to the parabolic MAN .

Let $\mathcal{H}$ be an irreducible Harish-Chandra $(\mathfrak{g}, K)$ module. We write $\mathcal{H} = \bigoplus_{\lambda \in \hat{K}} \mathcal{H}_\lambda$ where $\mathcal{H}_\lambda$ is the isotypic component of type λ .

Let us define the following subsets of $\overline{C}$:

6.1 Definition.

a) The K-support of $\mathcal{H}$, $S(\mathcal{H}) = \{\lambda \in \overline{C} \,;\, \mathcal{H}_\lambda \neq 0\}$.

b) The asymptotic K-support of $\mathcal{H}$

$$T(\mathcal{H}) = \{\lambda \in \overline{C} \text{ , such that there exists } \lambda_n \in S(\mathcal{H}) \text{ with } |\lambda_n| \to \infty \text{ , } t_n > 0 \text{ , and } t_n\lambda_n \to \lambda\}.$$

It is known that $\mathcal{H}$ can be imbedded as a $\mathfrak{g}$-submodule in a principal series $\operatorname{Ind}\uparrow_{MAN}^{G} \sigma \otimes \lambda$. Let us choose such an imbedding, and let us denote by H the completion of $\mathcal{H}$ in $L^2(X,U)$. Then G acts by bounded transformations on H. We denote by $SSH = \overline{\bigcup_{u\in H} SSu}$. Hence SSH is a closed subset of $\sqrt{-1}\, T^*K$. We identify $(SSH)_e$ as a subset of $\sqrt{-1}\, T_e^*K = \sqrt{-1}\, \mathfrak{k}^*$. We have $(Ad^*K)(SSH)_e = (Ad^*K)\, T(\mathcal{H})$, as follows from 5.5 and the proof of 5.4.

As H is stable by G, SSH is a G-invariant subset of $\sqrt{-1}\, T^*(K/M) \simeq \sqrt{-1}\, T^*(G/MAN)$. We have $T_0^*(G/MAN) \simeq (\mathfrak{m} + \mathfrak{a} + \mathfrak{n})^{\perp} \simeq \mathfrak{n}$ by the Killing form, $T_0^*(K/M) \simeq \mathfrak{m}^{\perp} \subset \mathfrak{k}^*$, the isomorphism $i: \mathfrak{n} \to \mathfrak{m}^{\perp}$ being given by the Killing form $X \longrightarrow B(X,\cdot)$. Hence if we identify $\mathfrak{k}^*$ with $\mathfrak{k}$, and we write $\mathfrak{g} = \mathfrak{k} \oplus \mathfrak{p}$, the map i is the restriction to $\mathfrak{n}$ of the orthogonal projection π from $\mathfrak{g}$ to $\mathfrak{k}$ perpendicular to $\mathfrak{p}$.

Let Q be an MAN invariant closed cone in $\mathfrak{n}$. Then SSH is of the form $\sqrt{-1}\, G\cdot Q = \sqrt{-1}\, K\cdot i(Q)$. We have:
$Ad^*K\cdot T(\mathcal{H}) = (Ad^*K)\cdot i(Q)$.

<u>6.2 Theorem</u>. Let S be a closed subset of nilpotent orbits of G in $\mathfrak{g}$. Let $\pi(S)$ be the projection of S to $\mathfrak{k}^*$ by the Killing form; let us denote by $T_S = \overline{C} \cap \sqrt{-1}\, \pi(S)$. For any Harish-Chandra module $\mathcal{H}$ there exists a closed subset S of nilpotent orbits of G in $\mathfrak{g}$ such that: $T(\mathcal{H}) = T_S$.

In particular for $\mathcal{H}$ a module of the principal series associated to MAN, we have $S = Ad^*G\cdot\mathfrak{n}$ = the nilpotent cone.

Let us give a example:

6.3 Example: $G = SU(2,1)$.

We consider the group $SU(2,1)$ associated to the canonical hermitian form

$$h = \begin{pmatrix} 1 & 0 & 0 \\ 0 & 1 & 0 \\ 0 & 0 & -1 \end{pmatrix}.$$

We choose

$$\mathfrak{a} = \begin{pmatrix} 0 & 0 & 1 \\ 0 & 0 & 0 \\ 1 & 0 & 0 \end{pmatrix}$$

and $K \simeq S(U(2) \times U(1))$. Then the group M is given by

$$\begin{pmatrix} e^{i\theta} & 0 & 0 \\ 0 & e^{-2i\theta} & 0 \\ 0 & 0 & e^{i\theta} \end{pmatrix}.$$

Hence $\sqrt{-1}\,\mathfrak{m}$ has basis

$$H_0 = \begin{pmatrix} 1 & 0 & 0 \\ 0 & -2 & 0 \\ 0 & 0 & 1 \end{pmatrix}.$$

Let $\mathfrak{h}$ be the Cartan subalgebra of $\mathfrak{k}$ given by

$$\sqrt{-1}\,\mathfrak{h} = \left\{ \begin{pmatrix} a_1 & 0 & 0 \\ 0 & a_2 & 0 \\ 0 & 0 & a_3 \end{pmatrix} ;\ a_1 + a_2 + a_3 = 0\ ,\ a_i \in \mathbb{R} \right\}.$$

We identify $\mathfrak{h}$ with its dual via the G-invariant form $(A,B) = \operatorname{Tr} AB$ for $A,B \in \mathfrak{su}(2,1)$. We identify the Lie algebra of $\sqrt{-1}\,\mathfrak{k}$ with the space of hermitian matrices by

$$A \to \begin{pmatrix} A & 0 \\ 0 & -\operatorname{Tr} A \end{pmatrix} :$$

then the orbits of K in $\sqrt{-1}\,\mathfrak{k}$ are classified by the eigenvalues of A . In this identification $\sqrt{-1}\,\mathfrak{m}^{\perp}$ is the subspace of matrices

$$\left\{ \begin{pmatrix} x_0 & u \\ \bar{u} & 0 \end{pmatrix} \right\}.$$

Hence an hermitian matrice

$$\begin{pmatrix} \lambda_1 & 0 \\ 0 & \lambda_2 \end{pmatrix}$$

is conjugated to $\mathfrak{m}^{\perp}$ if and only if $\lambda_1 \lambda_2 \leq 0$.

We have for $\Delta^+ = \{\alpha, \beta, \gamma\}$ the root system of $\mathfrak{g}$ with respect to $\mathfrak{h}$, α being the compact root, $\gamma = \beta + \alpha$.

$$H_\alpha = \begin{pmatrix} 1 & 0 & 0 \\ 0 & -1 & 0 \\ 0 & 0 & 0 \end{pmatrix}, \quad H_\beta = \begin{pmatrix} 0 & 0 & 0 \\ 0 & 1 & 0 \\ 0 & 0 & -1 \end{pmatrix}, \quad H_\gamma = H_\alpha + H_\beta .$$

The Weyl chamber C corresponding to the system of compact positive roots $\Delta_{\mathfrak{k}}^+ = \{\alpha\}$, is given by $\lambda(H_\alpha) > 0$. Hence $\bar{C} \cap \sqrt{-1}\,\mathfrak{h}_{\mathfrak{m}}^*$ is given by

$$\lambda = \{x_1 \beta + x_2 \gamma \quad x_1 \geq 0 \quad x_2 \leq 0\}$$

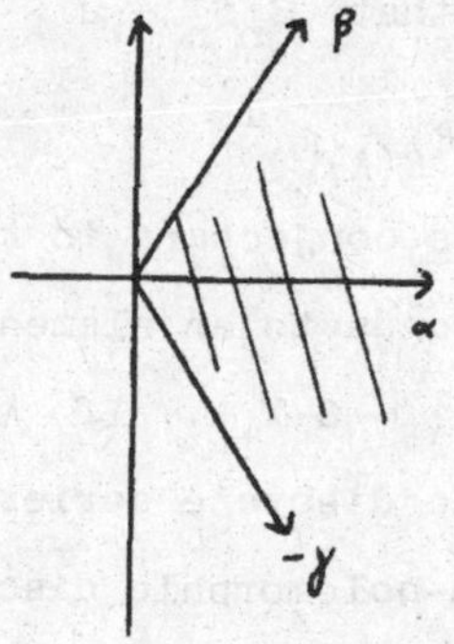

Let us consider the three possible classes of nilpotent elements for the action of G in $\mathfrak{g}$ ([6])

$$X_{\pm} = \pm \begin{pmatrix} 0 & 0 & 0 \\ 0 & i/2 & i/2 \\ 0 & -i/2 & -i/2 \end{pmatrix} , \quad X_0 = \begin{pmatrix} 0 & 1 & 0 \\ -1 & 0 & 0 \\ 1 & 0 & 0 \end{pmatrix} .$$

It is easily computed that

$$T_0 = \sqrt{-1}\, Ad^*G\cdot X_0 \cap \overline{C} = \sqrt{-1}\, \mathfrak{h}^*_{m} \cap \overline{C}$$

$$T_+ = \sqrt{-1}\, Ad^*G\cdot X_+ \cap \overline{C} = \mathbb{R}^+\cdot s \text{ , the half line of direction } s$$

$$T_- = \sqrt{-1}\, Ad^*G\cdot X_- \cap \overline{C} = -\mathbb{R}^+\cdot v \text{ , the half line of direction } v .$$

Let us precise our theorem (6.2) as follows: If the $(\mathfrak{g},K)$ module can be associated to an orbit Λ of Ad^*G in $\mathfrak{g}^*$ (for example, for the discrete series D_Λ, we will choose the G orbit of the elliptic element Λ) the choice of S should be given as follows: we define for an element $f \in \mathfrak{g}^*$ the asymptotic cone $S(f)$ to the orbit $G\cdot f$

i.e. $u \in S(f)$ if there exist $f_n \in G\cdot f$, $|f_n| \to \infty$ and $\varepsilon_n > 0$, such that $\varepsilon_n f_n \to u$

we then should have $T(\mathcal{X}_\Lambda) = T_{S(\Lambda)}$.[1]

It is easy to verify this conjecture in the case of $\mathfrak{g} = \mathfrak{su}(2,1)$: If Λ corresponds to an element of the holomorphic discrete series, we have $S(\Lambda) = G\cdot X_+$. If Λ corresponds to an element of the antiholomorphic discrete series, we have $S(\Lambda) = G\cdot X_-$. If Λ corresponds to the non-holomorphic discrete series, we have $S(\Lambda) = \overline{G\cdot X_0}$.

<u>Example 6.4</u>: Let $G = Sp(n,\mathbb{R})$ operating on the vector space $S(n)$ of symmetric $n \times n$ real matrices by $x \longrightarrow (ax + b)(cx + d)^{-1}$ for $\begin{pmatrix} a & b \\ c & d \end{pmatrix} \in Sp(n;\mathbb{R})$. The maximal compact subgroup K of G

[1]This conjecture has been proven recently by D. Barbasch and D. Vogan [9] for D_Λ.

is isomorphic to $U(n)$, via $a + ib \in U(n) \longrightarrow \begin{pmatrix} a & b \\ -b & a \end{pmatrix}$.

For P the parabolic

$$P = \left\{ \begin{pmatrix} a & 0 \\ * & {}^t a^{-1} \end{pmatrix} ; a \in GL(n;\mathbb{R}) \right\}$$

and μ a given finite dimensional representation μ of $GL(n;\mathbb{R})$, we consider the associated principal series $\operatorname{Ind}_P^G \mu = T_\mu$ (not necessarily unitary).

We denote by $M = P \cap K = O(n)$. We realize T_μ as a space of sections of a bundle over $G/P = K/M = X$. The vector space $S(n)$ can be considered as an open subset of G/P by $x \longrightarrow \begin{pmatrix} 1 & x \\ 0 & 1 \end{pmatrix}$ mod P , the corresponding action of G being given by the above formula. The corresponding identification

$$T^*(U(n)/O(n)) \simeq T^*S(n)$$

is given at the origin by $B \in S(n) \longrightarrow \begin{pmatrix} 0 & B \\ -B & 0 \end{pmatrix} \in \mathfrak{m}$. The pair $(K,M) = (U(n), O(n))$ is a symmetric pair. The preceding map allows us to identify the orthogonal complement of $\mathfrak{m}$ in $\mathfrak{k}$ with $S(n)$, the action of M on $\mathfrak{m}^\perp$ being given by $g \cdot X = gX{}^tg$, for $g \in O(n)$.

Let $\mathfrak{a}$ be the subspace of $\mathfrak{m}$ defined by diagonal matrices

$$\mathfrak{a} = \left\{ \begin{pmatrix} a_1 & & & \\ & a_2 & & \\ & & \ddots & \\ & & & a_n \end{pmatrix}, a_i \in \mathbb{R} \right\},$$

then every M-invariant subset of $\mathfrak{m}$ is of the form $M \cdot T$ where T is a subset of $\mathfrak{a}$.

$\mathfrak{a}$ is a Cartan subalgebra of $\mathfrak{k}$, hence every irreducible representation of K is indexed by its highest weight $\lambda = (\lambda_1, \lambda_2, \cdots, \lambda_n)$, where $\lambda_1 \geq \lambda_2 \geq \cdots \geq \lambda_n$, considered as an element of $\mathfrak{a}^* \simeq \mathfrak{a}$.

Let $\mathcal{H}$ be a $(\mathfrak{g},K)$ submodule of the space of K-finite vectors of the representation $\mathrm{Ind}\uparrow_P^G \mu$. We can analyze $SS\,H = \bigcup_{u\in H} SSu$ by analyzing the expansion of a function φ of H in terms of the K-Fourier series of $\varphi = \sum_{\lambda\in K} \varphi_\lambda$: i.e. let $\mathcal{H} = \bigoplus_{\lambda\in\hat{K}} \mathcal{H}_\lambda$, let $T(\mathcal{H})$ be the asymptotic K-support of $\mathcal{H}$ (definition 6.1). Let $M\cdot T(\mathcal{H}) \subset S(n)$ be the orbit of $T(\mathcal{H})$ under the group $O(n)$. We know that $SSH \subset T^*(K/M)$ is given by a K-invariant set of $T^*(K/M)$, and $(SSH)_e = M\cdot T(\mathcal{H})$, by 5.10.

Let us consider H as a G-module, then $SSH \subset T^*(G/P)$ is a G-invariant subspace of $T^*(G/P)$. Hence $(SSH)_e$ is given by a $GL(n,\mathbb{R})$ invariant closed subset of $S(n)$. The action of $GL(n)$ on $S(n)$ via $gX{}^tg$ decomposes $S(n)$ into a union of finite number of orbits $\mathcal{O}_{k_1,k_2}$, where $\mathcal{O}_{k_1,k_2}$ is the set of symmetric matrices of signature (k_1,k_2). Hence we have necessarily $(SSH)_e = \bigcup \overline{\mathcal{O}_{k_1,k_2}} \subset S(n)$ over a subset of orbits. Realizing H as a space of tempered distributions on the vector space $S(n)$, we may compute the singular spectrum of H using the Fourier integral $\hat{u}\ (\xi) = \int u(x)\, e^{-2i\pi \mathrm{Tr}\xi x}\, d\xi$ over the vector space $S(n)$ with respect to the bilinear form $\mathrm{Tr}\xi x$. If H is such that for every $u \in H$, $\hat{u}(\xi)$ is supported in $\bigcup\overline{\mathcal{O}_{k_1,k_2}}$, then $(SSH)_e \subset \bigcup\overline{\mathcal{O}_{k_1,k_2}}$.

Let us consider

$$\overline{C}_{k_1,k_2} = \{\lambda \in \overline{C}\ ;\ \lambda = (x_1,x_2,\cdots,x_i,0,\cdots,0,-y_j,-y_{j-1},\cdots -y_1)\},$$
$$\text{with } x_i \geq 0,\ y_j \geq 0,\ i \leq k_1,\ j \leq k_2\}.$$

We have $\overline{\mathcal{O}_{k_1,k_2}} = M\cdot\overline{C}_{k_1,k_2}$. Hence if H is a $(\mathfrak{g},K)$ module such that the asymptotic support of $\mathcal{H}$ is contained in a finite union of the sets C_{k_1,k_2}, it follows that $SSH \subset \overline{\mathcal{O}_{k_1,k_2}}$ and reciprocally.

This explains "asymptotically" the relation between the description of the spaces $H_{p,q}$ introduced in the article [7] via the support of the Fourier transform of the functions involved, and the K-support of $H_{p,q}$ given in [7].

In the similar example of the group U(2,2) acting by conformal transformations on the Minkowski space, we consider sub-representations H on the space of sections of the classical spin bundles on the Minkowski-space: We have in this case to consider $K = U(2) \times U(2)$, $M = U(2)$. Our bundles can be considered either as bundles over $K/M \simeq U(2)$, either on the flat Minkowski space identified with H(2) by

$$\vec{x} = (x_0, x_1, x_2, x_3) \longrightarrow x = \begin{pmatrix} x_0 + x_1 & x_2 + ix_3 \\ x_2 - ix_3 & x_0 - x_1 \end{pmatrix} .$$

The asymptotic directions of the K-types occurring in H are given by

$$T(H) = \{(m_1, m_2) \times (-m_2, -m_1) \in U(2)^\wedge \times U(2)^\wedge\} \text{ with } (m_1, m_2) \in T \subset \overline{C} \subset H(2)\}.$$

We can similarly read on the asymptotic directions of the K-types of H the support of the Fourier transform of a function u on H considered as a classical field. For example the space H of solutions of Maxwell, Dirac or Wave equation (considered as a subspace of the appropriate bundle) will have as asymptotic support the line $T = (m,0)$ as $U(2) \cdot T \subset H(2)$ is the light cone $(x_0^2 = x_1^2 + x_2^2 + x_3^2)$. The precise description of the support of H is given in [8].

References

[1] R. T. Seeley, Eigenfunction expansions of analytic functions, Proc. Amer. Math. Soc. 21, 1969, 734-738.

[2] Cerezo, A., Chazarain, J., Piriou A.: Introduction aux hyperfonctions. Lecture Notes in Math.

[3] Miwa, T., Oshima, T., Jimbo, M.: Introduction to micro-local analysis. Proceeding of the O.J.I. seminar on Algebraic Analysis. Publ. R.I.M.S. Kyoto Univ. 12 supplement, 267-300 (1966).

[4] Sato, M., Kawai, T., Kashiwara, M.: Microfunctions and pseudo differential equations. Lecture notes in Math. 287, pp. 265-529. Berlin, Heidelberg, New York: Springer 1973.

[5] King, D.: The geometric structure of the tensor product of irreducible representations of a complex semi-simple Lie algebra. Preprint, M.I.T. 1977.

[6] Barbasch, D.: Fourier inversion for unipotent invariant integrals, to appear in Trans. Amer. Math. Soc.

[7] Kashiwara, M., Vergne, M.: Functions on the Shilov boundary of the generalized half plane--Same volume.

[8] Speh, B.: Composition series for degenerate principal series representations of SU(2,2). Preprint, M.I.T. 1977.

[9] D. Barbasch and D. Vogan (to appear).

[10] A. Cerezo et F. Rouviere: Solution elementaire d'un operateur differentiel lineaire invariant a gauche sur un groupe de Lie riel compact et sur un espace homogene reductif compact. Ann. Sci. E.N.S. 4 (1969), 561-581.

A vanishing theorem for L^2-cohomology in the nilpotent case

Henri Moscovici

The motivation for studying realizations of the irreducible unitary representations of a nilpotent Lie group in terms of L^2-cohomology comes from Kostant's quantization theory, which aims to give a unified treatment of the construction of irreducible unitary representations of general Lie groups. The experience and, on the other hand, the analogy with the Borel-Weil-Bott theorem in the compact case or, more generally, with the Langlands realization of the discrete series representations in the semisimple noncompact case suggest the following behaviour of the L^2-cohomology spaces attached to a polarization of a nilpotent Lie algebra at a point of its dual vector space: they vanish in all but one dimension, and in the missing dimension, which is given by the signature of the corresponding nondegenerate symmetric bilinear form, the action of the group is irreducible. This is precisely the statement which has been verified in [3] under certain additional, perhaps unnecessary, conditions on the polarization. It is my purpose here to prove the vanishing part of this assertion under no restrictive assumptions on the polarization, but only for sufficiently large multiples of the given functional.

The result resembles the vanishing theorem of Griffiths and Schmid [2, Theorem 7.8] for "sufficiently nonsingular" weights in the semisimple case and, indeed, the proof is based on the same idea of getting information about the L^2-cohomology spaces by computing, as explicitly as possible, the Laplace-Beltrami operator. It must be said that in the nilpotent context the computation becomes simpler, for the obvious reason that one cannot expect to get a formula for the Laplace-Beltrami operator nearly so explicit as in the semisimple case.

To begin with, let me establish the notation, G will denote a connected and simply connected nilpotent Lie group with Lie algebra $\mathfrak{g}$. Given a linear functional λ on $\mathfrak{g}$, B_λ stands for the alternating bilinear form on $\mathfrak{g}$ given by

$$B_\lambda(x, y) = \lambda([x, y]), \qquad x, y \in \mathfrak{g}.$$

Its extension to the complexification $\mathfrak{g}_{\mathbb{C}}$ of $\mathfrak{g}$ will be denoted by the same symbo This convention will be also applied to the functional λ.

By $\mathfrak{h}$ I denote a polarization of $\mathfrak{g}$ at λ, that is a complex subalgebra of $\mathfrak{g}_{\mathbb{C}}$ which is a maximally isotropic subspace with respect to B_λ and has the further property that $\mathfrak{h} + \bar{\mathfrak{h}}$ is a subalgebra of $\mathfrak{g}_{\mathbb{C}}$. Set $\mathfrak{d} = \mathfrak{h} \cap \mathfrak{g}$ and $\mathfrak{e} = (\mathfrak{h} + \bar{\mathfrak{h}}) \cap \mathfrak{g}$; then $\mathfrak{d}_{\mathbb{C}} = \mathfrak{h} \cap \bar{\mathfrak{h}}$ and $\mathfrak{e}_{\mathbb{C}} = \mathfrak{h} + \bar{\mathfrak{h}}$. Further let D, E denote the connected (and, necessarily, simply connected) subgroups of G whose corresponding Lie algebras are $\mathfrak{d}$, $\mathfrak{e}$ respectively. The complexified tangent space of the homogeneous manifold $X = E/D$ at eD (e being the neutral element) can be identified with $(\mathfrak{e}/\mathfrak{d})_{\mathbb{C}} = \bar{\mathfrak{h}}/\mathfrak{d}_{\mathbb{C}} \oplus \mathfrak{h}/\mathfrak{d}_{\mathbb{C}}$, and X has precisely one E-invariant complex structure whose antiholomorphic tangent space at eD corresponds to $\mathfrak{h}/\mathfrak{d}_{\mathbb{C}}$.

Denote by ζ_λ the unitary character of D whose differential is $i\lambda|\mathfrak{d}$. Then ζ_λ associates to the principal bundle $D \to E \xrightarrow{p} X$ a complex line bundle $L_\lambda \to X$, whose space of C^∞-sections over an open subset U of X is canonically isomorphic to the space of C^∞-functions f on $p^{-1}(U)$ with the property:

$$f(ac) = \zeta_\lambda(c)^{-1} f(a), \quad a \in p^{-1}(U), \quad c \in D.$$

Further, L_λ can be turned into a holomorphic line bundle such that a function f as above defines a holomorphic section if and only if

$$uf + i\lambda(u)f = 0,$$

for every $u \in \mathfrak{h}$ extended as a left-invariant complex vector field on E. The space of all C^∞, L_λ-valued forms of type $(0, q)$ on X will be denoted by $\mathcal{A}^q(L_\lambda)$, and $\mathcal{A}_0^q(L_\lambda)$ will stand for the subspace of compactly supported forms in $\mathcal{A}^q(L_\lambda)$.

Obviously, L_λ carries an E-invariant Hermitian structure, which is unique up to multiplication by a constant. However to attach L^2-cohomology spaces to this line bundle one also needs an E-invariant Hermitian structure on X itself, more precisely on its complexified tangent bundle, and such a structure exists if and only if the image of Ad(D) in $\mathrm{End}(\mathfrak{e}/\mathfrak{d})$ is compact. If this is so, the polarization $\mathfrak{h}$ will be called a "metric polarization". Note that, in our nilpotent case, this amounts to the fact that $\mathfrak{d}$ is an ideal of $\mathfrak{e}$.

From now on $\mathfrak{h}$ will be always assumed a metric polarization. Once an E-invariant Hermitian structure on X is chosen, one may equip $\mathcal{A}_0^q(L_\lambda)$ with the E-invariant inner product arising from the E-invariant Hermitian metrics on L_λ and X by integration over X with respect to an E-invariant measure. Then the usual $\bar\partial$-operator $\bar\partial : \mathcal{A}^q(L_\lambda) \to \mathcal{A}^{q+1}(L_\lambda)$ has a formal adjoint $\bar\partial^* : \mathcal{A}_0^q(L_\lambda) \to \mathcal{A}_0^{q-1}(L_\lambda)$ and one can form the Laplace-Beltrami operator

$$\square = \bar\partial\bar\partial^* + \bar\partial^*\bar\partial .$$

It extends to an unbounded self-adjoint operator on the Hilbert space $\mathcal{L}_2^q(L_\lambda)$, which is by definition the completion of $\mathcal{A}_0^q(L_\lambda)$ with respect to the above mentioned inner product. Since the metric on X is necessarily complete and since $\square$ is elliptic, the kernel $\mathcal{H}^q(L_\lambda)$ of this extended Laplace-Beltrami operator consists exactly of those C^∞, L_λ-valued $(0, q)$-forms ω which satisfy the equations

$$\bar{\partial}\omega = 0 = \bar{\partial}^* \omega$$

in the ordinary sense. Now everything being E-invariant, E acts unitarily on the Hilbert space $\mathcal{H}^q(L_\lambda)$. The resulting unitary representation of E will be denoted $\pi^q(\lambda, \mathfrak{h}, E)$. Finally, as in [3], let

$$\pi^q(\lambda, \mathfrak{h}, G) = \mathrm{ind}_E^G \pi^q(\lambda, \mathfrak{h}, E)$$

and let $\mathcal{H}^q(\lambda, \mathfrak{h}, G)$ denote the corresponding Hilbert space.

To state the vanishing theorem I have in mind, one more notation is needed. The assignment

$$(u, v) \mapsto i\lambda([u, \bar{v}]), \qquad u, v \in \mathfrak{h},$$

gives rise to a nondegenerate sesquilinear form H_λ on $\mathfrak{h}/\mathfrak{d}_{\mathbb{C}}$. I shall denote by $q_{\lambda,\mathfrak{h}}$ the number of "negative squares" of H_λ.

THEOREM. There exists a positive number $t_{\lambda,\mathfrak{h}}$ such that $\mathcal{H}^q(t\lambda, \mathfrak{h}, G) = 0$ for all $t > t_{\lambda,\mathfrak{h}}$ and $q \neq q_{t\lambda,\mathfrak{h}} = q_{\lambda,\mathfrak{h}}$.

<u>Proof</u>. Obviously, one may assume from the beginning that G = E and $\lambda \neq 0$. Not only that but, since $[\mathfrak{e}, \mathfrak{d}] \subset \mathfrak{d} \cap \ker \lambda$, $\mathfrak{d} \cap \ker \lambda$ is an ideal of $\mathfrak{e}$, so that one may consider, without any loss of generality, only the situation $\mathfrak{d} \cap \ker \lambda = 0$. Then, $\lambda | \mathfrak{d}$ being necessarily nontrivial, $\mathfrak{d}$ is one-dimensional and, since it contains the center of $\mathfrak{e}$, it is precisely the center of $\mathfrak{e}$.

Set $\mathfrak{m} = \ker(\lambda | \mathfrak{h})$; $\mathfrak{m}$ is a Lie subalgebra, $\mathfrak{h} = \mathfrak{d}_{\mathbb{C}} \oplus \mathfrak{m}$, and

$$\mathfrak{e}_{\mathbb{C}} = \mathfrak{d}_{\mathbb{C}} \oplus \bar{\mathfrak{m}} \oplus \mathfrak{m}.$$

The complexified tangent space of X = E/D at eD can therefore be identified with $\bar{\mathfrak{m}} \oplus \mathfrak{m}$ and by this identification the antiholomorphic tangent subspace corresponds to $\mathfrak{m}$. Further, one may identify $\mathcal{A}^q(L_\lambda)$ with the subspace of those elements in $C^\infty(E) \otimes \Lambda^q \mathfrak{m}^*$ which transforms according to the character ζ_λ^{-1} when D is made to act on $C^\infty(E)$ by right translations and on $\mathfrak{m}^*$ by Ad^*. The latter action being trivial, $\mathcal{A}^q(L_\lambda)$ must be viewed in fact as $C^\infty(E, \zeta_\lambda) \otimes \Lambda^q \mathfrak{m}^*$, where $C^\infty(E, \zeta_\lambda)$ is formed by those $f \in C^\infty(E)$ such that

$$f(ac) = \zeta_\lambda(c)^{-1} f(a), \quad a \in E, \quad c \in D.$$

Similarly, $\mathcal{L}_2^q(L_\lambda)$ identifies to $L^2(E, \zeta_\lambda) \otimes \Lambda^q \mathfrak{m}^*$, where $L^2(E, \zeta_\lambda)$ stands for the completion of $C_0^\infty(E, \zeta_\lambda)$ with respect to the inner product

$$\langle f_1, f_2 \rangle = \int_{E/D} f_1(a)\overline{f_2(a)}\, d\dot{a}$$

($d\dot{a}$ means a left Haar measure on E/D) and $\mathfrak{m}$ is equipped with the inner product which comes from that we have chosen on $\mathfrak{e}_{\mathbb{C}}/\mathfrak{d}_{\mathbb{C}}$.

Once a basis $\{u_1, \ldots, u_n\}$ of $\mathfrak{m}$ is fixed, with dual basis $\{\omega^1, \ldots, \omega^n\}$ in $\mathfrak{m}^*$, the operators $\bar{\partial}$ and $\bar{\partial}^*$ get the following expressions:

$$\bar{\partial} = \Sigma_j\, u_j \otimes \varepsilon(\omega^j) + \tfrac{1}{2} \Sigma_j\, I \otimes \varepsilon(\omega^j) u_j,$$

$$\bar{\partial}^* = -\Sigma_j\, \bar{u}_j \otimes \iota(\omega^j) + \tfrac{1}{2} \Sigma_j\, I \otimes u_j^* \iota(\omega^j);$$

in these formulae u_j acts on $C_0^\infty(E, \zeta_\lambda)$ as a left-invariant complex vector field and on $\Lambda^q \mathfrak{m}^*$ as $\Lambda^q ad^*(u_j)$, u_j^* denotes the adjoint map of $\Lambda^q ad^*(u_j)$ with respect to the canonical inner product on $\Lambda^q \mathfrak{m}^*$, $\varepsilon(\omega^j)$ is the exterior multiplication by ω^j, and $\iota(\omega^j)$ denotes its adjoint operation.

In what follows it will be convenient to make a particular choice of the basis $\{u_1, \ldots, u_n\}$, such that the following conditions are fulfilled:

(CB) $H_\lambda(u_\alpha, u_\alpha) = 1$, $H_\lambda(u_\beta, u_\beta) = -1$ and $H_\lambda(u_j, u_k) = 0$,

where $\alpha = 1, \ldots, n - q_{\lambda,\mathcal{k}}$, $\beta = n - q_{\lambda,\mathcal{k}} + 1, \ldots, n$, and

$j \neq k$, $j, k = 1, \ldots, n$;

(OB) $\{u_1, \ldots, u_n\}$ is an orthonormal basis.

Such a choice is always possible, since one can first take a basis satisfying only (CB) and then give X the E-invariant Hermitian structure which arises from the inner product on $\mathfrak{m}$ uniquely determined by the condition (OB).

I want to compute now the Laplace-Beltrami operator $\Box$ acting on the space $\mathcal{A}^q(L_{t\lambda})$, where t is a fixed positive number. First of all let me introduc some useful terminological and notational conventions. By a "first order operator" I mean an operator of the form

$$\Sigma_j u_j \otimes \Phi_j + \Sigma_j \overline{u}_j \otimes \Psi_j + I \otimes \Omega,$$

where Φ_j, Ψ_j, $\Omega \in \mathrm{End}(\Lambda^q \mathfrak{m}^*)$, $j = 1, \ldots, n$.

I shall denote by $\overline{\partial}_1$, $\overline{\partial}_1^*$ the first "first order homogeneous part" of $\overline{\partial}$ and $\overline{\partial}^*$ respectively; more exactly

$$\overline{\partial}_1 = \Sigma_j u_j \otimes \varepsilon(\omega^j),$$

$$\overline{\partial}_1^* = -\Sigma_k \overline{u}_k \otimes \iota(\omega^k).$$

It is clear that

$$\Box = \Box_1 + F_1,$$

where

$$\Box_1 = \overline{\partial}_1 \overline{\partial}_1^* + \overline{\partial}_1^* \overline{\partial}_1$$

and F_1 is a first order operator.

Now

$$\square_1 = - \Sigma_{j,k} u_j \bar{u}_k \otimes \varepsilon(\omega^j)\iota(\omega^k) - \Sigma_{j,k} \bar{u}_k u_j \otimes \iota(\omega^k)\varepsilon(\omega^j)$$

$$= - \Sigma_j u_j \bar{u}_j \otimes \varepsilon(\omega^j)\iota(\omega^j) - \Sigma_j \bar{u}_j u_j \otimes \iota(\omega^j)\varepsilon(\omega^j)$$

$$- \Sigma_{j \neq k} u_j \bar{u}_k \otimes \varepsilon(\omega^j)\iota(\omega^k) - \Sigma_{j \neq k} \bar{u}_k u_j \otimes \iota(\omega^k)\varepsilon(\omega^j)$$

$$= - \Sigma_\alpha u_\alpha \bar{u}_\alpha \otimes \varepsilon(\omega^\alpha)\iota(\omega^\alpha) - \Sigma_\beta u_\beta \bar{u}_\beta \otimes \varepsilon(\omega^\beta)\iota(\omega^\beta)$$

$$- \Sigma_\alpha \bar{u}_\alpha u_\alpha \otimes \iota(\omega^\alpha)\varepsilon(\omega^\alpha) - \Sigma_\beta \bar{u}_\beta u_\beta \otimes \iota(\omega^\beta)\varepsilon(\omega^\beta)$$

$$- \Sigma_{j \neq k} u_j \bar{u}_k \otimes \varepsilon(\omega^j)\iota(\omega^k) - \Sigma_{j \neq k} \bar{u}_k u_j \otimes \iota(\omega^k)\varepsilon(\omega^j),$$

where α runs over the set $\Delta_+ = \{1, \ldots, n-q_{\lambda,\hbar}\}$ and β runs over $\Delta_- = \{n-q_{\lambda,\hbar} + 1, \ldots, n\}$. Using the identity

$$\varepsilon(\omega^j)\iota(\omega^k) + \iota(\omega^k)\varepsilon(\omega^j) = \delta^{jk}$$

it follows that

$$\square_1 = - \Sigma_\alpha [u_\alpha, \bar{u}_\alpha] \otimes \varepsilon(\omega^\alpha)\iota(\omega^\alpha) - \Sigma_\beta [\bar{u}_\beta, u_\beta] \otimes \iota(\omega^\beta)\varepsilon(\omega^\beta)$$

$$- \Sigma_\alpha \bar{u}_\alpha u_\alpha \otimes I - \Sigma_\beta u_\beta \bar{u}_\beta \otimes I$$

$$- \Sigma_{j \neq k} [u_j, \bar{u}_k] \otimes \varepsilon(\omega^j)\iota(\omega^k) .$$

Since $\mathfrak{e}_{\mathbb{C}} = \mathfrak{d}_{\mathbb{C}} \oplus \mathfrak{m} \oplus \bar{\mathfrak{m}}$, the brackets in the above formula can be expressed as

$$[u_j, \bar{u}_k] = c_{jk} z + \Sigma_i a^i_{jk} u_i + \Sigma_i b^i_{jk} \bar{u}_i ,$$

where z is a fixed element in $\mathfrak{d}$ with the property

$$\lambda(z) = 1.$$

Taking into account the fact that λ vanishes on $\mathfrak{m} \oplus \overline{\mathfrak{m}}$ and also the properties (CB), one finds that:

$$c_{\alpha\alpha} = -i, \qquad \text{for } \alpha \in \Delta_+ ;$$

$$c_{\beta\beta} = i, \qquad \text{for } \beta \in \Delta_- ;$$

$$c_{jk} = 0, \qquad \text{for } j \neq k.$$

Hence

$$\Box_1 = \Sigma_\alpha \, iz \otimes \varepsilon(\omega^\alpha)\iota(\omega^\alpha) + \Sigma_\beta \, iz \otimes \iota(\omega^\beta)\varepsilon(\omega^\beta)$$

$$- \Sigma_\alpha \overline{u}_\alpha u_\alpha \otimes I - \Sigma_\beta u_\beta \overline{u}_\beta \otimes I + F_2,$$

with F_2 a (homogeneous) first order operator.

But z being central, iz acts on $C^\infty(E, \zeta_{t\lambda})$ as the multiplication by t, so that

$$\Box_1 = t(\Sigma_\alpha \, I \otimes \varepsilon(\omega^\alpha)\iota(\omega^\alpha) + \Sigma_\beta \, I \otimes \iota(\omega^\beta)\varepsilon(\omega^\beta))$$

$$- \Sigma_\alpha \overline{u}_\alpha u_\alpha \otimes I - \Sigma_\beta u_\beta \overline{u}_\beta \otimes I + F_2,$$

and finally

$$\Box = t(\Sigma_\alpha \, I \otimes \varepsilon(\omega^\alpha)\iota(\omega^\alpha) + \Sigma_\beta \, I \otimes \iota(\omega^\beta)\varepsilon(\omega^\beta))$$

$$- \Sigma_\alpha \overline{u}_\alpha u_\alpha \otimes I - \Sigma_\beta u_\beta \overline{u}_\beta \otimes I + F,$$

where $F = F_1 + F_2$ is again a first order operator.

Given $J = (j_1, \ldots, j_q)$, an ordered q-tuple of integers between 1 and n, ω^J will stand for the exterior product $\omega^{j_1} \wedge \ldots \wedge \omega^{j_q}$. Every element in

$\mathcal{A}^q(L_{t\lambda})$ is of the form $\Sigma_J f_J \otimes \omega^J$, with J running over the set of all such q-tuples. Now fix an element $\varphi \in \mathcal{A}_0^q(L_{t\lambda})$,

$$\varphi = \Sigma_J f_J \otimes \omega^J,$$

where $f_J \in C_0^\infty(E, \zeta_{t\lambda})$. Then

$$\begin{aligned}
<\Box\varphi, \varphi> &= t\Sigma_{\alpha, J, K} <f_J, f_K><\varepsilon(\omega^\alpha)\iota(\omega^\alpha)\omega^J, \omega^K> \\
&\quad + t\Sigma_{\beta, J, K} <f_J, f_K><\iota(\omega^\beta)\varepsilon(\omega^\beta)\omega^J, \omega^K> \\
&\quad - \Sigma_{\alpha, J, K} <\bar{u}_\alpha u_\alpha f_J, f_K><\omega^J, \omega^K> \\
&\quad - \Sigma_{\beta, J, K} <u_\beta \bar{u}_\beta f_J, f_K><\omega^J, \omega^K> + <F\varphi, \varphi> \\
&= t(\Sigma_{\alpha, J} \|f_J\|^2 \|\iota(\omega^\alpha)\omega^J\|^2 + \Sigma_{\beta, J} \|f_J\|^2 \|\varepsilon(\omega^\beta)\omega^J\|^2) \\
&\quad + \Sigma_{\alpha, J} \|u_\alpha f_J\|^2 + \Sigma_{\beta, J} \|\bar{u}_\beta f_J\|^2 + <F\varphi, \varphi> \\
&= t\Sigma_J (\Sigma_{\alpha \in J} \|f_J\|^2 + \Sigma_{\beta \notin J} \|f_J\|^2) + \Sigma_{\alpha, J} \|u_\alpha f_J\|^2 \\
&\quad + \Sigma_{\beta, J} \|\bar{u}_\beta f_J\|^2 + <F\varphi, \varphi> \\
&= t\Sigma_J n_J \|f_J\|^2 + \Sigma_{\alpha, J} \|u_\alpha f_J\|^2 + \Sigma_{\beta, J} \|\bar{u}_\beta f_J\|^2 + <F\varphi, \varphi>,
\end{aligned}$$

where

$$n_J = \mathrm{card}(\Delta_+ \cap J) + \mathrm{card}(\Delta_- - J).$$

The point is that if $q \neq q_{\lambda, k}$, then $n_J \geq 1$ for every J. Therefore

$$<\Box\varphi, \varphi> \geq t\|\varphi\|^2 + \Sigma_{\alpha, J} \|u_\alpha f_J\|^2 + \Sigma_{\beta, J} \|\bar{u}_\beta f_J\|^2 - |<F\varphi, \varphi>|.$$

Noting that $|<u_j f_J, f_K>| = |<\bar{u}_j f_K, f_J>|$, it is easy to see that

$$|<F\varphi, \varphi>| \leq 2a\Sigma_{\alpha, J, K} |<u_\alpha f_J, f_K>| + 2a\Sigma_{\beta, J, K} |<\bar{u}_\beta f_J, f_K>| + c\|\varphi\|^2 .$$

Using this and the inequality $x^2 - 2mxy \geq -m^2y^2$, one finds that

$$\Sigma_{\alpha, J} \|u_\alpha f_J\|^2 - 2a\Sigma_{\alpha, J, K} |\langle u_\alpha f_J, f_K\rangle| \geq -(n-q_{\lambda,k})b^2\|\varphi\|^2$$

and similarly

$$\Sigma_{\beta, J} \|\bar{u}_\beta f_J\|^2 - 2a\Sigma_{\beta, J, K} |\langle \bar{u}_\beta f_J, f_K\rangle| \geq -q_{\lambda,k} b^2\|\varphi\|^2 ,$$

with $b = a\binom{n}{q}$.

Summing up, one finally obtains, for any $\varphi \in \mathcal{A}_0^q(L_{t\lambda})$,

$$\langle \Box\varphi, \varphi\rangle \geq (t-t_{\lambda,k})\|\varphi\|^2 ,$$

where $t_{\lambda,k} = nb^2 + c$. In view of [1, Proposition 8], this concludes the proof.

References

1. Andreotti, A., Vesentini, E.: Carleman estimates for the Laplace-Beltrami operator on complex manifolds. Publ. I. H. E. S., 25, 81-130 (1965).

2. Griffiths, P., Schmid, W.: Locally homogeneous complex manifolds. Acta Mathematica, 123, 253-302 (1969).

3. Moscovici, H., Verona, A.: Harmonically induced representations of nilpotent Lie groups. Inventiones mathematicae (to appear).

Permanent address: INCREST, Dept. of Mathematics
Bd. Păcii 220, Bucharest 77538

Current address: The Institute for Advanced Study
Princeton, NJ 08540

The Eichler Commutation Relation and the Continuous Spectrum of the Weil Representation

by

Stephen Rallis

Introduction

The object of this lecture is to give a representation-theoretic proof of the Eichler Commutation Relation in the theory of θ-series. In qualitative terms, this relation can be stated as follows. If $\mathfrak{M}$ is the linear space of θ-series attached to the distinct classes in a fixed genus of a positive definite, integral quadratic form Q (defined on a $2m$ dimensional space), then $\mathfrak{M}$ is stable under all Hecke operators $T(m)$, where m is the norm of a similarity transform in the similitude group $S(Q)$ of Q.

It is possible to interpret this relation in terms of the local Weil representation. In particular, the observation we make is that the determination of the continuous spectrum of the local Weil representation completely specifies which Hecke operators on $S(Q)$ correspond to Hecke operators on $G\ell_2$ relative to the Eichler Commutation Relation. We construct a certain homomorphism Cor_p of the local Hecke algebra of $S(Q)$ to the local Hecke algebra of $G\ell_2$. Then we show that the Eichler Commutation Relation can be given in terms of a lifting from automorphic forms on $S(Q)$ to automorphic forms on $G\ell_2$ (in an adelic setting) which commutes with the Cor_p homomorphism (Theorem 2.1).

Knowing explicitly the Cor_p homomorphism (given in §4), it is then possible to deduce (in certain cases) a version of Siegel's formula in the analytic theory of positive definite quadratic forms (Remark 1.1 and §5). We note that the proof is representation-theoretic. That is, we characterize the lift to $G\ell_2$ of the identity automorphic representation of $S(Q)$ in terms of the eigenvalues of the $G\ell_2$ Hecke algebras acting on the local components of the lift.

Table of Contents

§1. The Classical Theory

We consider the classical version of the Eichler Commutation Relation in the theory of θ-series.

We let $\mathbb{R}^{2m}$ be a Euclidean 2m-dimensional space. We choose once and for all a basis $\{e_i\}_{i=1}^{2m}$ on $\mathbb{R}^{2m}$. Then let L_0 be the $\mathbb{Z}$-lattice in $\mathbb{R}^{2m}$ given by the $\mathbb{Z}$ span of the vectors $\{e_i\}$. Let $\mathbb{Q}^{2m}$ be the rational subspace of $\mathbb{R}^{2m}$ given by the $\mathbb{Q}$-span of the $\{e_i\}$. Also let $[\ ,\ ]$ be the dot product on $\mathbb{R}^{2m}$ satisfying $[e_i,e_j] = \delta_{ij}$ for all i,j.

Let $\{\ \}$ be the set of all rational lattices in $\mathbb{Q}^{2m}$. That is, $L \in \{\ \}$ is a free $\mathbb{Z}$-submodule of $\mathbb{Q}^{2m}$ of rank 2m. Then since the lattice L is a finitely generated $\mathbb{Z}$-module, the set $\{(1/2)[\xi,\xi] \mid \xi \in L\}$ generates a fractional ideal in $\mathbb{Q}$. Thus there exists a positive rational number $n(L)$, the norm of L, so that $n(L)$ generates this fractional ideal over $\mathbb{Z}$. We say that $L \in \{\ \}$ is <u>even integral</u> if $n(L)$ is an integer.

We consider the <u>dual lattice</u> L^x to L given by $L^x = \{\mu \in \mathbb{Q}^{2m} \mid \frac{1}{n(L)}[\mu,L] \subseteq \mathbb{Z}\}$. Then $L^x \in \{\ \}$ and $L^x \supseteq L$.

Then, to a given $L \in \{\ \}$, we associate an equivalence class of integral symmetric matrices on $\mathbb{R}^{2m}$. First we note that $G\ell_{2m}(\mathbb{Z})$ operates on the $\mathbb{Z}$-space S_m of all integral symmetric matrices (with <u>even</u> diagonal elements) by $M \in S_m \rightsquigarrow {}^tUMU$ with $U \in G\ell_{2m}(\mathbb{Z})$. Then, given $L \in \{\ \}$, we let $\{\xi_i\}_{i=1}^{i=2m}$ be a $\mathbb{Z}$-basis of L. We then consider the integral symmetric matrix $S_L = \frac{1}{n(L)}\mathrm{mat}([\xi_i,\xi_j])$. Another choice of basis of L gives an integral symmetric matrix which is equivalent to S_L via the action of $G\ell_{2m}(\mathbb{Z})$ discussed above. Then we say that two lattices L_1 and L_2 belong to the <u>same</u> $G\ell_{2m}(\mathbb{Z})$ <u>class</u> if S_{L_1} and S_{L_2} are $G\ell_{2m}(\mathbb{Z})$ <u>equivalent</u>.

We let $q(L)$ be the smallest positive integer so that $q(L)S_L^{-1}$ belongs to S_m. Then it is easy to see that $q(L)S_L^{-1}$ belongs to the $G\ell_{2m}(\mathbb{Z})$ equivalence class of S_L (i.e. $q(L) = \frac{n(L)}{n(L^x)}$).

Moreover we let $\det(S_L)$ be the determinant of the matrix S_L. Then it is easy to see that $\varepsilon_L(x) = (\frac{(-1)^m \det(S_L)}{x})(\operatorname{sgn} x)^m$ is independent of the choice of basis of L (here $(\frac{*}{x})$ is the Legendre symbol). Also $x \to \varepsilon_L(x)$ determines a character on the group $(\mathbb{Z}/q(L)\mathbb{Z})^x$ (= the invertible elements in the ring $\mathbb{Z}/q(L)\mathbb{Z}$).

Then for any $L \in \{\ \}$, we can define a θ-series in the following way:

$$(1.1) \qquad \theta_L(z) = \sum_{\xi \in L} e^{\pi\sqrt{-1}\, z \frac{1}{n(L)}[\xi,\xi]} = \sum_{n \geq 0} A_L(n) e^{\pi\sqrt{-1}\, nz},$$

where $A_L(n) = \{\xi \in L \mid [\xi,\xi] = n \cdot n(L)\}$ is the classical representation number of n by L. Then one knows that the function $\theta_L \in [\Gamma_0(q(L)), \varepsilon_L, m] = \{\varphi\colon H \to \mathbb{C} \mid \varphi \text{ holomorphic}, \varphi(\gamma \cdot z) = \varepsilon_L(d_\gamma)(c_\gamma z + d_\gamma)^m \varphi(z)$ for all $\gamma \in \Gamma_0(q(L))$, and φ is "regular" at the cusps of $\Gamma_0(q(L))$ on $H \cup \mathbb{Q}\}$, where $H = \{z = x + \sqrt{-1}\, y \mid y > 0\}$ is the upper half plane and $\Gamma_0(q(L)) = \{\gamma \in SL_2(\mathbb{Z}) \mid c_\gamma \equiv 0 \bmod q(L)\}$ is the Hecke congruence group of level $q(L)$ (here "regular" at cusps is taken in the sense of [12]).

We recall the action of a Hecke operator $T(n)$ on the space $[\Gamma_0(q), \varepsilon, m]$ (here ε is a fixed character on $(\mathbb{Z}/q\mathbb{Z})^x$ to $\mathbb{Z}_2 = \{\pm 1\}$). That is, $T(n)$ is given by

$$(1.2) \qquad T(n)f(z) = n^{m-1} \sum_{\substack{ad=n \\ b \bmod d \\ d>0}} \varepsilon(a) d^{-m} f(\frac{az+b}{d}),$$

with $f \in [\Gamma_0(q),\varepsilon,m]$. Then $[\Gamma_0(q),\varepsilon,m]$ is stable under the family of operators $T(n)$ for $n \geq 1$. Moreover if W is any nonzero subspace of $[\Gamma_0(q),\varepsilon,m]$ stable under $T(n)$ for all $(n,q) = 1$, then W has a basis f_i of functions which are simultaneously eigenfunctions for all the $T(n)$ with $(n,q) = 1$, i.e. $T(n)f_i = \lambda(n)f_i$ for all i.

One of the main problems in the analytic theory of quadratic forms is to see what multiplicative properties the representation numbers $A_L(n)$ have. In particular, this reduces to the very difficult question of how the Hecke operators $T(n)$ operate on θ-series θ_L in general. To get a precise arithmetic statement about the exact nature of the $A_L(n)$ seems to be untractable at the present. However, what is reasonable is to get some qualitative statement about the stability of certain types of θ-series under the Hecke operators. That is, we consider a space of the form $\mathcal{A}_X$ = {the complex linear span of $\theta_L(z)$} where L ranges over representatives of a certain finite subset X of $G\ell_{2m}(\mathbb{Z})$ equivalence classes in [ ]. The main quesion is what conditions on X are needed so that $\mathcal{A}_X$ is stable under a suitable family of $T(n)$.

Remark 1.1. The first striking example of a linear combination of θ series stable under Hecke operators is given by the Theorem of Siegel in the analytic theory of quadratic forms. We assume here that $q = 1$. Let $L_1,\ldots,L_t$ be representatives of $G\ell_{2m}(\mathbb{Z})$ equivalence classes of even integral unimodular lattices (in this case $4|m$). Then let ε_{L_i} equal the cardinality of the finite group $\{g \in G\ell_{2m}(\mathbb{R}) \mid g(L_i) = L_i$ and g is orthogonal relative to [,]}. We then form the function $h(z) = \sum_i \frac{1}{\varepsilon_{L_i}} \theta_{L_i}(z)$.

Then Siegel's Theorem asserts that $h(z)$ is (up to a nonzero scalar) an Eisenstein series on H relative to the group $SL_2(\mathbb{Z})$; hence h is an eigenfunction for all $T(n)$ (recall that $q = 1$).

However, with the exception of the example above, no precise information can be obtained about the functions g_i. But there is yet another possible approach set forth in the work of Eichler [4]. Indeed we want to look more carefully at the arithmetic of lattices in $\{\ \}$ relative to the form $[\ ,\ ]$ on $\mathbb{Q}^{2m}$.

First we must recall the notion of a <u>genus</u> of a lattice. We let $\mathcal{O}_p$ be the ring of integers in the p-adic field $\mathbb{Q}_p$ (the p-adic completion of Q). We let $\mathfrak{p}$ be the maximal ideal in $\mathcal{O}_p$. Then for any lattice $L \in \{\ \}$, we see that $\mathcal{O}_p \otimes L = L_p$ is an $\mathcal{O}_p$ module of rank 2m in $\mathbb{Q}_p^{2m}$; hence L_p is a lattice in $\mathbb{Q}_p^{2m}$. Then we know that every lattice $L \in \{\ \}$ is determined by its local components L_p for all p finite, that is, $L = \mathbb{Q}^{2m} \cap (\bigcap_{p<\infty} L_p)$. From this remark, we see that $G\ell_{2m}(\mathbb{A})$ has an action on the set $\{\ \}$ (here $G\ell_{2m}(\mathbb{A})$ is the adelic group associated to $G\ell_{2m}(\mathbb{Q})$ and $\mathbb{A}$, the rational adeles). In particular if $g \in G\ell_{2m}(\mathbb{A})$, then $g = \prod g_p$ acts on L in the following fashion: $g(L) = \mathbb{Q}^{2m} \cap (\bigcap_{p<\infty} g_p(L_p))$ (note that $g_p(L_p) = L_p$ for almost all p). Then $G\ell_{2m}(\mathbb{A})$ acts transitively on the set $\{\ \}$.

We let S be the connected component of $\{g \in G\ell_{2m}(\mathbb{R}) \mid [gX, gX] = \lambda(g)[X,X]$ for all $X \in \mathbb{R}^{2m}\}$. S is the connected group of similitudes of the form $[\ ,\ ]$. Then since $[\ ,\ ]$ is defined over $\mathbb{Q}$, it is possible to form the corresponding adelic group $S_{\mathbb{A}}$. Then $S_{\mathbb{A}}$ is a closed subgroup of $G\ell_{2m}(\mathbb{A})$. We can then restrict the action of $G\ell_{2m}(\mathbb{A})$ on $\{\ \}$ to the group $S_{\mathbb{A}}$. However, $S_{\mathbb{A}}$ does not operate transitively on $\{\ \}$; we say that the $S_{\mathbb{A}}$ orbit of an $L \in \{\ \}$ is the S <u>genus</u> of L.

Then we note that one of the deeper facts in the reduction theory of lattices is that the S genus of L has, at most, a finite number of $G\ell_{2m}(\mathbb{Z})$ classes. It is possible to explicitly parameterize these classes. That is, we let $U_L = \{g \in S_{\mathbb{A}} | g(L) = L\}$. Then we note the standard embedding of $S_{\mathbb{Q}}$ into $S_{\mathbb{A}}$. Then there exists a bijection between the $G\ell_{2m}(\mathbb{Z})$ classes in the S genus of L and the double cosets in the $S_{\mathbb{Q}}\backslash S_{\mathbb{A}}/U_L$. That is, each class in the S genus of L consists of all lattices of the form $S_{\mathbb{Q}}\alpha(L)$, where $S_{\mathbb{Q}}\alpha U_L$ is a distinct double coset in $S_{\mathbb{Q}}\backslash S_{\mathbb{A}}/U_L$. We let $M_1,\ldots,M_h$ be the representatives of the $G\ell_{2m}(\mathbb{Z})$ classes in the S genus of L. Then we form the space $\mathfrak{M}(\mathrm{gen}(L),q(L),\varepsilon_L) = \{$complex linear span of $\theta_{M_i}(z)$, $i = 1,\ldots,h\}$. Then $\mathfrak{M}(\mathrm{gen}(L),q(L),\varepsilon_L) \subseteq [\Gamma_0(q(L)),\varepsilon_L,m]$ from the comments above.

Theorem 1.1. The space $\mathfrak{M}(\mathrm{gen}(L),q(L),\varepsilon_L)$ is stable under all $T(n)$ where $(n,2q(L)) = 1$ and n is the norm of a similarity transform associated to S_L (i.e. there exists a rational matrix x so that $x^t S_L x = n \cdot S_L$).

The theorem of Eichler is remarkable in that it also provides an explicit way to determine the effect of $T(p)$ on the space $\mathfrak{M}(\mathrm{gen}(L),q(L),\varepsilon_L)$. We assume that $m \geq 2$. First we note that p, a prime, must satisfy the hypotheses of Theorem 1.1. We let $\tilde{\varepsilon}_{M_i}$ = the order of the finite group $\{g \in S | g(M_i) = M_i$ and $|\lambda(g)| = 1\}$. Let $W = \mathrm{diag}(\tilde{\varepsilon}_{M_i})$.

The conditions on p given above imply that for each M in the S genus of L, the form $[\ ,\ ]$ induces on the quotient $M/pM \cong M_p/pM_p$ (which is a vector space of dimension $2m$ over the finite field $\mathfrak{O}_p/p\mathfrak{O}_p \cong \mathbb{F}_p$) a nondegenerate quadratic form on $\mathbb{F}_p^{2m}$ which is equivalent to an orthogonal direct sum of m hyperbolic planes V (i.e. $V = \langle x\rangle \perp \langle y\rangle$ with $[x,x] = [y,y] = 0$ and $[x,y] = 1$). Then we fix a pair M_i and M_j of distinct

$G\ell_{2m}(\mathbb{Z})$ classes in the S genus of L. Then we let $c_{ij}(p)$ = the number of lattices M in the S genus of L which have the properties

(i) $(M_j) \supseteq M \supseteq p\, M_j$ and $M/p\, M_j$ determines a <u>maximal</u> $[\ ,\ ]$ isotropic subspace in $\mathbb{F}_p^{2m}$,

(ii) M is $S_{\mathbb{Q}}$ conjugate to M_i, i.e. there exists $\gamma \in S_{\mathbb{Q}}$ so that $\gamma(M) = M_i$.

Then we consider the matrix $A(p) = \mathrm{mat}(c_{ij}(p))$. The effect of $T(p)$ on the space $\mathfrak{M}(\mathrm{gen}(L), q(L), \epsilon_L)$ then can be expressed in the following way:

$$(1.3) \qquad T(p) * \begin{pmatrix} \theta_{M_1}(\ ) \\ \cdot \\ \vdots \\ \theta_{M_h}(\ ) \end{pmatrix} = d(p)\, WA(p)W^{-1} \begin{pmatrix} \theta_{M_1}(\ ) \\ \cdot \\ \vdots \\ \theta_{M_h}(\ ) \end{pmatrix},$$

where $d(p) = \frac{w_p}{\deg(A(p))}$ with $\deg(A(p))$ = the number of maximal isotropic subspaces in $\mathbb{F}_p^{2m}$, and w_p, the eigenvalue of $T(p)$ on the modular form $\sum \frac{1}{\tilde{\epsilon}_{M_i}} \theta_{M_i}$ (we show in <u>Remark</u> 5.1 that this fact is true!).

§2. The Eichler Lifting

<u>We assume for the rest of this paper that</u> $m \geq 2$.

Our goal is to generalize <u>Theorem</u> 1.1 in an adelic setup and to give a representation theoretic proof of this result. The first step is to interpret the space $[\Gamma_0(q), \epsilon, m]$ in the adelic language. But this is fairly well known. We recall that $G\ell_2(A)$ can be decomposed as $K^q_{\mathrm{fin}} \cdot G\ell_2^+(\mathbb{R}) \cdot G\ell_2(\mathbb{Q})$, where $K^q_{\mathrm{fin}} = \prod_{p<\infty} K^q_p$ with $K^q_p = \{\gamma \in G\ell_2(\mathcal{O}_p) \mid c_\gamma \equiv 0 \bmod q\}$,

$G\ell_2^+(\mathbb{R}) = \{g \in G\ell_2(\mathbb{R}) \mid \det g > 0\}$, and $G\ell_2(\mathbb{Q})$ is embedded in $G\ell_2(\mathbb{A})$ in the standard fashion.

Given any Dirichlet character on $(\mathbb{Z}/q\mathbb{Z})^x$, such as ε, it is possible to construct a corresponding character ε_p on $\mathcal{O}_p^x$ via the natural homomorphism of $\mathcal{O}_p^x$ to $(\mathbb{Z}/p^m\mathbb{Z})^x$ (where m = the p-adic order of q). Then we can define a character on $\prod_{p<\infty} \mathcal{O}_p^x$ by taking $\prod_{p<\infty} \varepsilon_p$; in fact, this character extends to a grossencharacter ε_* on I, the <u>ideles</u> in A, i.e. we use the fact that $I = \mathbb{Q}^x \cdot \mathbb{R}_\infty^+ \prod_{p<\infty} \mathcal{O}_p^x$ and then extend $\prod_{p<\infty} \varepsilon_p$ so that it is trivial on $\mathbb{Q}^x \cdot \mathbb{R}_\infty^+$. Then we can formally extend ε_* to a character on the group K_{fin}^q as follows: $\prod k_p \rightsquigarrow \prod_{p<\infty} \varepsilon_p(a_p)$, where $k_p = \begin{pmatrix} a_p & b_p \\ c_p & d_p \end{pmatrix} \in K_p^q$ for all $p < \infty$.

On the other hand, we recall that the factor of automorphy on H relative to the group $G\ell_2^+(\mathbb{R})$ is $j(g,z) = \det(g)^{-1/2} \cdot (c_g z + d_g)$. We know that j satisfies a well known cocycle condition, i.e. $j(g_1 g_2, z) = j(g_1, g_2 \cdot z) j(g_2, z)$ for all $g_1, g_2 \in G\ell_2^+(\mathbb{R})$ and $z \in H$.

Thus, to each element $f \in [\Gamma_0(q), \varepsilon, m]$, we can associate a function ϕ_f on $G\ell_2(\mathbb{A})$ given by: $\phi_f(g) = f(g_\infty^{-1}(\sqrt{-1}))j(g_\infty^{-1}, \sqrt{-1})^{-m}\, \varepsilon_*(k_0)$, where $g \in G\ell_2(\mathbb{A})$ can be written as $g = k_0 g_\infty \gamma$. Then an easy exercise shows that $\phi_f(k_1 r(\theta) g\gamma) = \varepsilon_*(k_1) e^{\sqrt{-1}\theta m} \phi_f(g)$ for $k_1 \in K_{fin}^q$, $r(\theta) = \begin{pmatrix} \cos\theta & -\sin\theta \\ \sin\theta & \cos\theta \end{pmatrix} \in G\ell_2^+(\mathbb{R})$ and $\gamma \in G\ell_2(\mathbb{Q})$. Then we denote {the complex linear span on ϕ_f as f varies in $[\Gamma_0(q), \varepsilon, m]$} by $[\Gamma_0(q), \varepsilon, m]^*$. we note that the elements in $[\Gamma_0(q), \varepsilon, m]^*$ also satisfy certain growth conditions on $G\ell_2(\mathbb{A})$ and a certain differential equation on the infinite component $G\ell_2^+(\mathbb{R})$ (for a precise statement of these conditions, see [6, p. 42]).

Then for p a prime so that $p \nmid q$, we consider the local Hecke algebra $\mathcal{H}(G\ell_2)(\mathbb{Q}_p)//K_p^q)$ = {the linear span of all compactly supported functions f on $G\ell_2(\mathbb{Q}_p)$ which satisfy $\varpi(k_1 g k_2) = \varpi(g)$ for all $k_1, k_2 \in K_p^q\}$, where $K_p^q = G\ell_2(\mathcal{O}_p)$. We note that, once having fixed a Haar measure dx on $G\ell_2(\mathbb{Q}_p)$, the algebra $\mathcal{H}(G\ell_2(\mathbb{Q}_p)//K_p^q)$ is closed under convolution; in fact, it is a commutative algebra. Then it is possible to define an action of the algebra $\mathcal{H}(G\ell_2(\mathbb{Q}_p)//K_p^q)$ on $[\Gamma_0(q),\epsilon,m]^*$ by convolution on the left, i.e. if $\varpi \in \mathcal{H}(G\ell_2(\mathbb{Q}_p)//K_p^q)$ and $F \in [\Gamma_0(q),\epsilon,m]^*$, then

$$(2.1) \qquad \varpi * F(g) = \int_{G\ell_2(\mathbb{Q}_p)} \varpi(y)\, F(y^{-1}g)\, dy$$

defines an element in $[\Gamma_0(q),\epsilon,m]^*$ (assuming here that $\int_{G\ell_2(\mathcal{O}_p)} dx = 1$).

Then, in particular, we note that if X_p is the characteristic function of $G\ell_2(\mathcal{O}_p)\begin{pmatrix} p & 0 \\ 0 & 1 \end{pmatrix}G\ell_2(\mathcal{O}_p)$, then $X_p * \phi_f = p^{1-(m/2)}\phi_{T(p)f}$. Also if Z_p is the characteristic function of $G\ell_2(\mathcal{O}_p)\begin{pmatrix} p & 0 \\ 0 & p \end{pmatrix}G\ell_2(\mathcal{O}_p)$, then $Z_p * \phi_f = \phi_f$.

We now fix a lattice L in $\{\ \}$. Then for some $g \in G\ell_{2m}(\mathbb{Q})$, we have that $g(L_0) = L$ (recall L_0 is the standard lattice in $\mathbb{Q}^{2m}$). We let $S_L = \frac{1}{n(L)}\, g^t g$. Then we consider the connected group $\tilde{S}$ of similitudes of S_L, i.e. $\tilde{S}$ is the connected component of $\{h \in G\ell_{2m}(\mathbb{R})\ |h^t S_L h = \nu(h) S_L\}$. Then $\tilde{S}$ is clearly defined over $\mathbb{Q}$, and we consider the adelic group $\tilde{S}_{\mathbb{A}}$. Moreover, since $g^{-1} S_{\mathbb{A}} g = \tilde{S}_{\mathbb{A}}$ in $G\ell_{2m}(\mathbb{A})$, it follows that there is a bijective correspondence between the $G\ell_{2m}(\mathbb{Z})$ classes in the $\tilde{S}_{\mathbb{A}}$ genus of L_0 and the $G\ell_{2m}(\mathbb{Z})$ classes in the $S_{\mathbb{A}}$ genus of L (i.e. $S_{\mathbb{A}}(L) = g\tilde{S}_{\mathbb{A}}(L_0)$). Moreover, we let $\tilde{S}^1_{\mathbb{A}} = \{h \in \tilde{S}_{\mathbb{A}} | \prod_p |\nu_p(h)|_p = 1\}$, where ν_p is the similitude factor given by the local group $\tilde{S}_p$ = the connected component of $\{h \in G\ell_{2m}(\mathbb{Q}_p) | h^t (S_L)_p h = \nu_p(h)(S_L)_p\}$, with $(S_L)_p$ viewed as an element in $M_{2m}(\mathbb{Q}_p)$ ($2m \times 2m$ matrices of $\mathbb{Q}_p$). Then $\tilde{S}^1_{\mathbb{A}}$ is closed subgroup of $\tilde{S}_{\mathbb{A}}$. We know that $\tilde{S}_{\mathbb{A}}$ is the direct product of $\tilde{S}^1_{\mathbb{A}}$ and Z_∞, where Z_∞

is the closed subgroup of $\tilde{S}_\infty$ given by $Z_\infty = \{\lambda \cdot I \mid \lambda \in \mathbb{R}$ and $\lambda > 0\}$. Moreover, we have that $\tilde{S}_{\mathbb{Q}} \subset \tilde{S}^1_{\mathbb{A}}$ and that the homogeneous space $\tilde{S}^1_{\mathbb{A}}/\tilde{S}_{\mathbb{Q}}$ is compact. We let $\tilde{U}_{L_0} = \{h \in \tilde{S}_{\mathbb{A}} \mid h(L_0) = L_0\}$; then $\tilde{U}_{L_0} \cap \tilde{S}^1_{\mathbb{A}} = U'_{L_0}$ is the product $\prod_{p<\infty} U'_p \times \tilde{S}^1_\infty$ where $U'_p = \{x \in \tilde{S}_p \mid x((L_0)_p) = (L_0)_p\}$ and $\tilde{S}^1_\infty = \{x \in \tilde{S}_\infty \mid |\nu_\infty(x)| = 1\}$. In particular, we see that each U'_p $(p < \infty)$ and $\tilde{S}^1_\infty$ are compact subgroups of $\tilde{S}_p$ and $\tilde{S}_\infty$, respectively. Moreover, it is a straightforward exercise to verify that the double cosets in $\tilde{U}_{L_0} \backslash \tilde{S}_{\mathbb{A}}/\tilde{S}_{\mathbb{Q}}$ can be put into one-one correspondence with the double cosets in $U'_{L_0} \backslash \tilde{S}^1_{\mathbb{A}}/\tilde{S}_{\mathbb{Q}}$.

Then we consider the space $F(U'_{L_0} \backslash \tilde{S}^1_{\mathbb{A}}/\tilde{S}_{\mathbb{Q}}) = \{\varphi\colon \tilde{S}^1_{\mathbb{A}} \to \mathbb{C} \mid \varphi$ measurable function and $\varphi(x_1 g x_2) = \varphi(g)$ for $x_1 \in U'_{L_0}$ and $x_2 \in \tilde{S}_{\mathbb{Q}}$ and $g \in \tilde{S}^1_{\mathbb{A}}\}$. Then $F(U'_{L_0} \backslash \tilde{S}^1_{\mathbb{A}}/\tilde{S}_{\mathbb{Q}})$ is a finite dimensional vector spacc spanned by the characteristic functions of the double cosets in $U'_{L_0} \backslash \tilde{S}^1_{\mathbb{A}}/\tilde{S}_{\mathbb{Q}}$. Moreover, from the comments above, $F(U'_{L_0} \backslash \tilde{S}^1_{\mathbb{A}}/\tilde{S}_{\mathbb{Q}})$ is isomorphic in the obvious way to the space $F(U'_{L_0} \backslash \tilde{S}_{\mathbb{A}}/\tilde{S}_{\mathbb{Q}} Z_\infty) = \{\varphi\colon \tilde{S}_{\mathbb{A}} \to \mathbb{C} \mid \varphi$ measurable and $\varphi(\mathfrak{z}_1 g \mathfrak{z}_2) = \varphi(g)$ for all $\mathfrak{z}_1 \in U'_{L_0}$, $\mathfrak{z}_2 \in \tilde{S}_{\mathbb{Q}} \cdot Z_\infty$, and $g \in \tilde{S}_{\mathbb{A}}\}$.

Then since U'_{L_0} is a compact subgroup of $\tilde{S}_{\mathbb{A}}$, we form the global Hecke algebra $\mathcal{H}(\tilde{S}_{\mathbb{A}}//U'_{L_0}) = \{f\colon \tilde{S}_{\mathbb{A}} \to \mathbb{C} \mid \varphi$ is continuous and compactly supported and $f(x_1 g x_2) = f(g)$ for all $x_1, x_2 \in U'_{L_0}\}$. Then let dx be the Tamagawa measure on $\tilde{S}_{\mathbb{A}}$ constructed in the manner of [1]; we know that $\mathcal{H}(\tilde{S}_{\mathbb{A}}//U'_{L_0})$ is a commutative algebra under convolution. Then we define the local Hecke algebras $\mathcal{H}(\tilde{S}_p//U'_p) = \{\varphi : \tilde{S}_p \to \mathbb{C} \mid \varphi$ is smooth and compactly supported on $\tilde{S}_p$ and $\varphi(\ell_1 g \ell_2) = \varphi(g)$ with $\ell_1, \ell_2 \in U'_p\}$ ($\mathcal{H}(\tilde{S}_\infty // \tilde{S}_\infty^1)$ is defined similarly). Then we consider the relevant convolution structures on $\mathcal{H}(\tilde{S}_p // U'_p)$ and $\mathcal{H}(\tilde{S}_\infty // \tilde{S}_\infty^1)$ (where the choice of local Haar measures dx_p and dx_∞ on $\tilde{S}_p$ and $\tilde{S}_\infty$, respectively, is such that $\int_{U'_p} dx_p = 1$ and $\int_{\tilde{S}_\infty^1} dx_\infty = 1$). We note that the local measure dx_p does not necessarily coincide with the local measure on $\tilde{S}_p$ used to define the Tamagawa measure on $\tilde{S}_{/A}$. Then we know that there exists an algebra embedding of $\mathcal{H}(\tilde{S}_p // U'_p)$ ($\mathcal{H}(\tilde{S}_\infty // \tilde{S}_\infty^1)$, resp.) into $\mathcal{H}(\tilde{S}_{/A} // U'_{L_0})$ given by the map

$\varphi \overset{i_p}{\rightsquigarrow} \frac{1}{\mathrm{vol}_{dx}(U'_{L_0})} \varphi \otimes \prod_{\{q | q \neq p\}} \chi_q$ (where χ_q is the characteristic function of U'_q if $q \neq \infty$ and the characteristic function of $\tilde{S}_\infty^1$ if $q = \infty$).

Then the algebra $\mathcal{H}(\tilde{S}_{/A} // U'_{L_0})$ has a natural representation on $F(U'_{L_0} \backslash \tilde{S}_A^1 / \tilde{S}_{\mathbb{Q}})$ given by convolution on the left, i.e.

$$(2.2) \qquad f * F_1(g) = \int_{\tilde{S}_{/A}} f(x)\, F_1(x^{-1} g)\, dx ,$$

with $f \in \mathcal{H}(\tilde{S}_{/A} // U'_{L_0})$ and $F_1 \in F(U'_{L_0} \backslash \tilde{S}_{/A}^1 / \tilde{S}_{\mathbb{Q}}) \cong F(U'_{L_0} \backslash \tilde{S}_{/A} / \tilde{S}_{\mathbb{Q}} Z_\infty)$. It is easy to describe the action of $f *$ on the space $F(U'_{L_0} \backslash \tilde{S}_{/A} / \tilde{S}_{\mathbb{Q}})$. Indeed, we let ρ_i be the characteristic function of $U'_{L_0} \alpha_i \tilde{S}_{\mathbb{Q}}$ (so that $\bigcup_i U'_{L_0} \alpha_i \tilde{S}_{\mathbb{Q}}$ is a disjoint union of $\tilde{S}_{/A}^1$). Then we can write

$$(2.3) \qquad f * \rho_i = \sum_j c_{ij}(f)\, \rho_j ,$$

where

(2.4) $$c_{ij}(f) = \int_{\widetilde{S}_A} f(x)\rho_i(x^{-1}\alpha_j)\, dx \ .$$

Let f be the characteristic function of the set of the form $U'_{L_0}\beta U'_{L_0}$ with $\beta = \beta_{fin}\cdot\beta_\infty$, where $\beta_\infty = 1$. One knows that $U'_{L_0}\beta U'_{L_0}$ consists of a finite number of left cosets $\gamma_k^\beta U'_{L_0}$. Then we have that $c_{ij}(f) = \text{volume}_{dx}(U'_{L_0}\beta U_{L_0} \cap \alpha_j \widetilde{S}_{\mathbb{Q}} \alpha_i^{-1} U'_{L_0}) = \text{volume}_{dx}(U'_{L_0}) \cdot$ {the number of cosets $\gamma_k^\beta U'_{L_0}$ which can be represented by a coset of the form $\alpha_j \xi \alpha_i^{-1} U'_{L_0}$ with $\xi \in \widetilde{S}_{\mathbb{Q}}$}.

<u>Remark</u> 2.1. Let p be a prime so that $p \nmid 2q(L)$. The one knows that $(L_0)_p$ is a maximal lattice in $\mathbb{Q}_p^{2m}$ (that is, $(L_0)_p$ is maximal among all $\mathfrak{O}_p$-lattices in $\mathbb{Q}_p^{2m}$ having the same norm $n((L_0)_p) = \mathfrak{O}_p$). Then if we assume that $\beta = \beta_{fin}$ has only the p-th component nontrivial and that $\beta((L_0)_p) \subseteq (L_0)_p$, then $\beta_p(L_0)_p$ is a maximal $\mathfrak{O}_p$-lattice with norm equal to $\nu_p(\beta_p)n((L_0)_p) = \mathfrak{p}^{\ell'}$. Then from the theory of elementary divisors and the local arithmetic of quadratic forms, we know that there exists a set $\{e_i, e'_j, (1 \le i \le \nu)\}$ in $\mathbb{Q}_p^{2m}$ so that $(L_0)_p = \sum_{i=1}^{i=\nu} \mathfrak{O}_p e_i \oplus \sum_{i=1}^{i=\nu} \mathfrak{O}_p e'_i \oplus L_0^{(0)}$ and $\beta_p((L_0)_p) = \sum_{i=1}^{i=\nu} \mathfrak{p}^{r_i} e_i \oplus \sum_{i=1}^{i=\nu} \mathfrak{p}^{\ell'-r_i} e_i \oplus L_0^{(\ell')}$, with $r_1 \ge r_2 \ge \cdots \ge r_\nu \ge 1/2\ \ell'$, and $L_0^{(0)} (L_0^{(\ell')}$ resp.) is the unique $\mathfrak{O}_p$ maximal lattice in $(\sum_{i=1}^{i=\nu} \mathbb{Q}_p e_i \oplus \sum_{i=1}^{i=\nu} \mathbb{Q}_p e'_i)^\perp$ ($\perp$ denoting orthogonal complement) of the smallest possible exponential norm ≥ 0 (resp. $\ge \ell'$). We note there that $\langle e_i, e_j\rangle = \langle e'_i, e'_j\rangle = 0$, $\langle e_i, e'_j\rangle = \delta_{ij}$ relative to the bilinear form $\langle\ ,\ \rangle$ induced

on $\mathbb{Q}_p^{2m}$ by the matrix $(S_L)_p$. Then we know that the ordered set of integers $(r_1, r_2, \ldots, r_\nu)$ is uniquely determined, independent of the choice of the set $\{e_i, e_i'\}$. We call $(r_1, \ldots, r_\nu)$ the set of <u>elementary divisors</u> of $\beta_p((L_0)_p)$ relative to $(L_0)_p$.

We then know that if $\nu = m-1 < m$, then the set of all maximal $\mathfrak{O}_p$ lattices M' in $\mathbb{Q}_p^{2m}$ having norm $\rho^{\ell'}$ so that $M' \subseteq (L_0)_p$ and M' has elementary divisors $(r_1, \ldots, r_\nu)$ relative to $(L_0)_p$ is in one-one correspondence with the set of left U_p' cosets in $U_p' \beta_p U_p'$. On the other hand if $\nu = m$, then the set of $\mathfrak{O}_p$-lattices with the above conditions is in one-one correspondence with the left cosets in the disjoint union $U_p' \beta_p U_p'$ and $U_p' \beta_p^\nu U_p'$ if $r_m \neq 1/2\ \ell'$ or $U_p' \beta_p U_p'$ if $r_m = 1/2\ \ell'$ (where $\beta_p^\nu = w_m \beta_p w_m^{-1}$ with w_m the similitude which is the identity of $\langle e_m, \tilde{e}_m \rangle^\perp$ and $w_m(e_m) = \tilde{e}_m, w_m(\tilde{e}_m) = e_m$).

Thus it is now possible to give an arithmetic interpretation of the number $c_{ij}(U_{L_0}' \beta U_{L_0}')$ $(c_{ij}(U_{L_0}' \beta U_{L_0}') + c_{ij}(U_{L_0}' \beta^\nu U_{L_0}')$, resp.), where $\beta = \beta_p$ satisfying the hypotheses in the above Remark (where $\beta_p^\nu = w_m \beta_p w_m^{-1}$ is defined above in the case that $\nu = m$ and $r_m \neq 1/2\ \ell'$). In particular, we consider the set of all lattices R' in $\{\ \}$ so that

(i) R' is contained in the $\mathfrak{S}_A$ genus of L_0

(ii) $\alpha_j(R')_p \subseteq (L_0)_p$, $\alpha_j(R')_p$ has $(r_1, \ldots, r_\nu)$ as a system of

(2.3) <u>elementary divisors</u> relative to $(L_0)_p$, and

$$\frac{n((L_j)_p)}{n((R')_p)} = \rho^{-\ell'}$$

(iii) R' can be conjugated by an element of $\mathfrak{S}_\mathbb{Q}$ to L_i.

Here L_i is the $G\ell_{2m}(\mathbb{Z})$ class representative corresponding to the double coset $\mathfrak{S}_\mathbb{Q}\, \alpha_i^{-1} U_{L_0}'$, i.e. $\alpha_i^{-1}(L_0) = L_i$. Then $c_{ij}(U_{L_0}' \beta U_{L_0}')$ $(c_{ij}(U_{L_0}' \beta U_{L_0}') +$

$c_{ij}(U'_{L_0}\beta^{\nu}U'_{L_0})$ if $\nu = m$ and $\beta = \beta_p$ has elementary divisors $(r_1,\dots,r_m)$ satisfying $r_m \neq 1/2\,\ell'$) equals the product of $\mathrm{volume}_{dx}(U'_{L_0})$ times the number of lattices R' in $\{\ \}$ satisfying (i), (ii), and (iii) above. We note that in the example given at the end of §1, the system of elementary divisors is $(\overbrace{1,1,\dots,1}^{m})$ and $\dfrac{n((L_j)_p)}{n((R')_p)} = p^{-1}$.

We are now in a position to define the Eichler lifting map. That is, we define a linear map $\mathcal{L} : F(U'_{L_0}\backslash\widetilde{S}^1_{A}/\widetilde{S}_{Q}) \to \mathcal{M}(\mathrm{gen}(L), q(L), \varepsilon_L)^{x}$. In particular, for the characteristic function ρ_i of the set $U'_{L_0}\alpha_i\widetilde{S}_{Q}$, we define $\mathcal{L}(\rho_i) = \frac{1}{\tilde{\varepsilon}_{L_i}}\,\Phi_{\theta_{L_i}}$, where $\tilde{\varepsilon}_{L_i}$ = the order of the <u>finite group</u> $\{g \in \widetilde{S}^1_{\infty} \mid g(L_i) = L_i\}$. We note that $\Phi_{\theta_{L_i}}$ is independent of the choice of the lattice in the $G\ell_{2m}(\mathbb{Z})$ class containing L_i. Then we extend $\mathcal{L}$ linearly to the space $F(U'_{L_0}\backslash\widetilde{S}^1_{A}/\widetilde{S}_{Q})$.

Now we can state the main Theorem of this paper, generalizing <u>Theorem 1.1</u>. We recall here that the local Haar measures on $\widetilde{S}_p$ and $G\ell_2(\mathbb{Q}_p)$ (used in defining the convolution structure on $\mathcal{H}(\widetilde{S}_p//U'_p)$ and $\mathcal{H}(G\ell_2(\mathbb{Q}_p)//G\ell_2(\mathcal{O}_p))$ are such that both U'_p and $G\ell_2(\mathcal{O}_p)$ have mass equal to 1 (relative to these measures).

<u>Theorem</u> 2.1. <u>Let</u> p <u>be a prime so that</u> $(p,2q(L)) = 1$. <u>Then there exists a homomorphism</u>:

(2.4) $$\mathrm{Cor}_p : \mathcal{H}(\widetilde{S}_p//U'_p) \to \mathcal{H}(G\ell_2(\mathbb{Q}_p)//G\ell_2(\mathcal{O}_p))$$

<u>so that</u>

(2.5) $$\mathcal{L}(i_p(\varpi) \times f) = \mathrm{Cor}_p(\check{\varpi}) * \mathcal{L}(f)$$

for $f \in F(U'_{L_0} \backslash \tilde{S}_{\mathbb{A}} / \tilde{S}_{\mathbb{Q}})$, where $\check{\varphi}$ denotes the element given by $\check{\varphi}(x) = \varphi(x^{-1})$ with $\varphi \in \mathcal{H}(\tilde{S}_p // U'_p)$. We need here the condition that $(\frac{(-1)^m}{p}) = 1$ if $(S_L)_p$ is totally split at $\mathbb{Q}_p$, i.e. as an orthogonal direct sum of hyperbolic planes.

We shall give an outline of the proof of Theorem 2.1 in §3. Also we shall give the Cor_p homomorphism explicitly in §4.

Remark 2.2. The condition that $(\frac{(-1)^m}{p}) = 1$ when $(S_L)_p$ is totally split at $\mathbb{Q}_p$ is actually too restrictive in the above Theorem. Indeed, if $(S_L)_p$ is totally split, then we can define a homomorphism Cor'_p of $\mathcal{H}(\tilde{S}_p // U'_p)'$ to $\mathcal{H}(G\ell_2(\mathbb{Q}_p) // G\ell_2(\mathcal{O}_p))$ so that $\mathcal{L}(i_p, (\varphi) * f) = Cor'_p(\check{\varphi}) * \mathcal{L}(f)$ is valid where $\mathcal{H}(\tilde{S}_p // U'_p) = \{\varphi \in \mathcal{H}(\tilde{S}_p // U'_p) \mid \text{support } (\varphi) \subseteq \{x \in \tilde{S}_p \mid \nu_p(x) \in (k^x)^2\}\}$

§3. The Weil Representation and the Lifting Problem

The first step in the proof of Theorem 2.1 is to describe the map $\mathcal{L}$ by means of an integral kernel operator. For this we need to introduce the global Weil representation attached to S_L.

We recall that locally for every place $\{p\}$ of $\mathbb{Q}$ (including $\{\infty\}$), there exists a unitary representation $\pi_p : \tilde{S}_p \times G\ell_2(\mathbb{Q}_p) \to \text{Unit}\,(L^2(\mathbb{Q}_p^{2m} \times \mathbb{Q}_p^x, dx \otimes dt))$ called the Weil representation (for an explicit description, we refer the reader to [6], [7], or [10]). In any case, it is then possible to construct the "smooth" global Weil representation $\pi_{S_L} = \pi$ of $\tilde{S}_{\mathbb{A}} \times G\ell_2(\mathbb{A})$ of $\mathcal{S}(\mathbb{A}^{2m} \times I)$ Schwartz-Bruhat functions (again an explicit description is given in [6]), where I is the group of ideles in $\mathbb{A}$. The key property of this representation is the existence of the Poisson distribution,

$\Theta = \sum_{\substack{\xi \in \mathbb{Q}^{2m} \\ \eta \in \mathbb{Q}^{x}}} \delta_{(\xi,\eta)}$ (where $\delta_{(\xi,\eta)}$ is the Delta distribution supported at (ξ,η) in $\mathbb{A}^{2m} \times I$), which has the feature that $\langle \Theta, \pi(\gamma_1,\gamma_2)\varphi \rangle = \langle \Theta, \varphi \rangle$ for all $(\gamma_1,\gamma_2) \in \tilde{S}_{\mathbb{Q}} \times G\ell_2(\mathbb{Q})$ and all $\varphi \in \mathcal{S}(\mathbb{A}^{2m} \times I)$. This means that the map

$\varphi \rightsquigarrow \Theta_{\varphi}(g_1,g_2) = \sum_{\substack{\xi \in \mathbb{Q}^{2m} \\ \eta \in I}} \pi(g_1,g_2)^{-1}(\varphi)(\xi,\eta)$ is an $\tilde{S}_{\mathbb{A}} \times G\ell_2(\mathbb{A})$ intertwining map from π to the space of measurable functions on $\tilde{S}_{\mathbb{A}} \times G\ell_2(\mathbb{A})$ which are invariant on the right by $\tilde{S}_{\mathbb{Q}} \times G\ell_2(\mathbb{Q})$. Thus it is possible to define the integral (which is absolutely convergent)

$$(3.1) \qquad \int_{\tilde{S}^1_{\mathbb{A}}/\tilde{S}_{\mathbb{Q}}} f(g_1)\Theta_{\varphi}(g_1,g_2)\, dg_1$$

for $f \in F(U'_{L_0} \backslash \tilde{S}^1_{\mathbb{A}}/\tilde{S}_{\mathbb{Q}})$ (where the measure dg_1 is the quotient measure on the compact homogeneous space $\tilde{S}^1_{\mathbb{A}}/\tilde{S}_{\mathbb{Q}}$ constructed from the measures on $\tilde{S}^1_{\mathbb{A}}$ and $\tilde{S}_{\mathbb{Q}}$ given by (i) the induced measure on $\tilde{S}^1_{\mathbb{A}}$ coming from the measure dx on $\tilde{S}_{\mathbb{A}}$ given above and (ii) the usual counting measure on $\tilde{S}_{\mathbb{Q}}$). Then in order to relate this integral to the map $\mathcal{L}$, we must make a judicious choice of the function $\varphi \in \mathcal{S}(\mathbb{A}^{2m} \times I)$. In particular, we let $\varphi = \varphi_0 = \prod_{p<\infty} \varphi_p \times \varphi_\infty$, where $\varphi_p = \chi_{(L_0)_p} \otimes \chi_{\frac{1}{n((L_0)_p)}\mathcal{O}_p^x}$ (with $\chi_{(L_0)_p}$, the characteristic function of the lattice $(L_0)_p$ in $\mathbb{Q}_p^{2m}$ and $\chi_{\frac{1}{n((L_0)_p)}\mathcal{O}_p^x}$, the characteristic function of $\frac{1}{n((L_0)_p)}\mathcal{O}_p^x$ in $\mathbb{Q}_p^x$) and $\varphi_\infty(Z,t) = e^{-\pi|t|_+ Z^t S_L Z}$ (with $Z \in \mathbb{R}^{2m}$, $t \in \mathbb{R}^x$, and $|t|_+ = \begin{cases} t & \text{if } t > 0 \\ 0 & \text{if } t < 0 \end{cases}$).
Then φ_0 transforms according to the character $\pi(k_1 r(\theta))\varphi_0 = \epsilon_*(k_1)e^{-\sqrt{-1}\, m\theta}\varphi_0$, where $k_1 \in K^q_{fin}$ and $r(\theta) = \begin{pmatrix} \cos\theta & -\sin\theta \\ \sin\theta & \cos\theta \end{pmatrix} \in G\ell_2^+(\mathbb{R})$. Thus we have that

$\Theta_{\varphi_0}(g_1, k_1 r(\theta) g_2) = \epsilon_*(k_1) e^{\sqrt{-1}\, m\theta}\, \Theta_{\varphi_0}(g_1, g_2).$

Lemma 3.1. Let $f \in F(U'_{L_0} \backslash \tilde{S}^1_{\mathbb{A}} / \tilde{S}_{\mathbb{Q}})$. Then

$$(3.2) \qquad \mathfrak{L}(f)(h) = \frac{1}{\mathrm{vol}_{dx}(U'_{L_0})} \int_{\tilde{S}^1_{\mathbb{A}}/\tilde{S}_{\mathbb{Q}}} f(g)\, \Theta_{\varphi_0}(g,h)\, dg \ .$$

Proof. It suffices to take $f = \rho_i$. Then an easy computation shows that $\int_{\tilde{S}^1_{\mathbb{A}}/\tilde{S}_{\mathbb{Q}}} \rho_i(g) \Theta_{\varphi_0}(g,h)\, dg = \mathrm{vol}_{dg}(U'_{L_0} \alpha_i \tilde{S}_{\mathbb{Q}}/\tilde{S}_{\mathbb{Q}}) \theta_{L_i}(h)$, where $\mathrm{vol}_{dg}(\ )$ is taken relative to the induced measure dx on $\tilde{S}^1_{\mathbb{A}}/\tilde{S}_{\mathbb{Q}}$ given above. Then following the arguments in [9, p. 262], one can deduce that $\mathrm{vol}_{dg} = (U'_{L_0} \alpha_i \tilde{S}_{\mathbb{Q}}/\tilde{S}_{\mathbb{Q}})$ $\mathrm{vol}_{dx}(U'_{L_0}) \frac{1}{e_{L_i}}$. Q.E.D.

Then using Lemma 3.1, it is now convenient to apply a Hecke operator $\varphi \in \mathcal{H}(\tilde{S}_{\mathbb{A}}//U'_{L_0})$ to $f \in F(U'_{L_0} \backslash \tilde{S}^1_{\mathbb{A}}/\tilde{S}_{\mathbb{Q}})$. We then find that $\mathfrak{L}(\varphi * f)(h) = \frac{1}{\mathrm{vol}_{dx}(U'_{L_0})} \langle \varphi * f | \Theta_{\varphi_0}(\ ,h)\rangle$, where $\langle f_1, f_2\rangle$ denotes the bilinear form on $\tilde{S}^1_{\mathbb{A}}/\tilde{S}_{\mathbb{Q}}$ given by $\int_{\tilde{S}^1_{\mathbb{A}}/\tilde{S}_{\mathbb{Q}}} f_1(g) f_2(g)\, dg$. Then we have that $\langle \varphi * f | \Theta_{\varphi_0}(\ ,h)\rangle = \langle f | \check{\varphi} * \Theta_{\varphi_0}(\ ,h)\rangle$, where $\check{\varphi}(x) = \varphi(x^{-1})$. (Note that $\check{\varphi}\, *$ is relative to the $\tilde{S}^1_{\mathbb{A}}$ variable). On the other hand, using the intertwining property of Θ_{φ_0}, we have that $\check{\varphi} * \Theta_{\varphi_0}(g,h) = \Theta_{r(\check{\varphi})(\varphi_0)}(g,h)$.

Thus the problem becomes purely a local matter. That is, we must determine the effect of $\pi_v(\varphi_v)$ on φ_p. In effect, we want to find a Hecke operator $\mathrm{Cor}_p(\varphi_v)$ in $\mathcal{H}(G\ell_2(\mathbb{Q}_p)//K^q_p)$ so that $(\pi_v(\varphi_v) - \pi_v(\mathrm{Cor}_p(\varphi_v)))(\varphi_p) \equiv 0$. This is precisely the technique used by Eichler in §11 of [4] for certain Hecke operators in $\mathcal{H}(\tilde{S}_p//U'_p)$ (we mean here that, by hindsight, one interprets Eichler's result in terms of the local Weil representation). The main disadvantage of this method is that it involves, in general, rather complicated arithmetic computations in the local theory of lattices.

We present a method for the proof of Theorem 2.1 which is representation theoretic, i.e. does not depend on counting techniques. Namely, we consider the "smooth" local Weil representation π_p of $\tilde{S}_p \times G\ell_2(\mathbb{Q}_p)$ on the space $S(\mathbb{Q}_p^{2m} \times \mathbb{Q}_p^x)$ of local Schwartz Bruhat functions (for the precise formulation, see [7]). We let $H_p = \mathcal{H}(\tilde{S}_p//U_p') \otimes \mathcal{H}(G\ell_2(\mathbb{Q}_p)//G\ell_2(\mathcal{O}_p))$ be the Hecke algebra of $\tilde{S}_p \times G\ell_2(\mathbb{Q}_p)$ relative to the maximal compact subgroup $U_p' \times G\ell_2(\mathcal{O}_p)$. Then, in any case, the problem that we consider is to find an explicit set of generators for the ideal

(3.3) $$I_p = \{\psi \in H_p \mid \pi_p(\psi) \equiv 0\}.$$

We call the set of generators of I_p the set of relations of the local Weil representation π_p.

Remark 3.1. We note that in [8], it is shown that $\pi_p(\mathcal{H}(\tilde{S}_p//U_p'))$ and $\pi_p(\mathcal{H}(G\ell_2(\mathbb{Q}_p)//G\ell_2(\mathcal{O}_p))$ generate the same image when operating on the space $S(\mathbb{Q}_p^{2m} \times \mathbb{Q}_p^x)$. What is not explicit in [8] is how the correspondence between Hecke operators is actually given.

The first specific question in this program is then to determine the continuous spectrum of π_p (for $p \nmid 2q(L)$). We then have that $(S_L)_p$ on $\mathbb{Q}_p^{2m}$ has the following form: there exists a set $\{w_i, \tilde{w}_i\}_{i=1}^{i=\nu}$ in $\mathbb{Q}_p^{2m}$ so that $(S_L)_p \cong \sum_{i=1}^{i=\nu} \mathbb{Q}_p w_i \oplus \sum_{i=1}^{i=\nu} \mathbb{Q}_p \tilde{w}_i \oplus E$, where $\langle w_i, w_j\rangle = \langle \tilde{w}_i, \tilde{w}_j\rangle = 0$ and $\langle w_i, \tilde{w}_j\rangle = \delta_{ij}$ (with $\langle\ ,\ \rangle$ the associated bilinear form on $\mathbb{Q}_p^{2m}$ constructed from $(S_L)_p$) and $\langle\ ,\ \rangle$ defines a nondegenerate anisotropic form on E equal either to $\{0\}$ or to a nonzero scalar times the norm form of the unique, unramified, quadratic extension of $\mathbb{Q}_p$. (Note $\nu = m-1$ if $\dim E > 0$ and $\nu = m$ if $\dim E = 0$.)

Then we fix a minimal parabolic P in $\widetilde{S}_p$ of the following type. We let $A_i(t,\lambda)$ be the subgroup of $\widetilde{S}_p$ which operates by $\begin{cases} w_i \to tw_i \\ \widetilde{w}_i \to t^{-1}\lambda\widetilde{w}_i \end{cases}$ $(t,\lambda \in k^x)$ and as identity on $\langle w_i, \widetilde{w}_i\rangle^{\perp}$. Then a maximal $\mathbb{Q}_p$ split torus in $\widetilde{S}_p$ is given by the group $A \cong \prod_{i=1}^{i=\nu} A_i(\xi_i, \xi^{[2]}) \cdot (\xi \cdot I_E) \cong (\prod_{i=1}^{i=\nu} \mathbb{Q}_p^x) \times \mathbb{Q}_p^x$, when I_E is the identity operator on E and

$\xi^{[2]} = \begin{cases} \xi & \text{if } \dim E = 0 \\ \xi^2 & \text{if } \dim E = 2 \end{cases}$. Then $P = Z(A) \cdot U$, where U is the unipotent radical of P, and $Z(A)$ is the centralizer of A in $\widetilde{S}_p$; it is easy to deduce that $Z(A)$ is isomorphic to the group $(\prod_{i=1}^{i=\nu} (\mathbb{Q}_p^x)) \times G_0$, where G_0 is $\mathbb{Q}_p^x$ (if $\dim E = 0$) or is the connected component of the group of similitudes of E (if $\dim E > 0$) with the isomorphism given by $(\xi_1,\ldots,\xi_\nu,g) \rightsquigarrow \prod_{i=1}^{i=\nu} A_i(\xi_i,\nu_p(g)) \cdot \begin{cases} I & \text{if } \dim E = 0 \\ g & \text{if } \dim E > 0 \end{cases}$ (where $\nu_p(g) = \xi$ if $g = \xi \in G_0 = \mathbb{Q}_p^x$). We note that the explicit form of U is not necessary for our purposes here (see [10] for more details). Then an easy exercise shows that the admissible, irreducible representations of P which are trivial on U are of the form:

(3.4) $$(\xi_1,\ldots,\xi_\nu,g) \overset{\chi}{\rightsquigarrow} \prod_{i=1}^{i=\nu} \chi_i(\xi_i) \otimes \chi_{\nu+1}(\nu_p(g)) \otimes \sigma(g) ,$$

where σ is an admissible, irreducible (finite dimensional) representation of G_0 which does not factor through the ν_p map, and $\{\chi_i\}$ $i = 1,\ldots,\nu+1$

is a set of quasicharacters on $\mathbb{Q}_p^x$.

On the other hand, a minimal parabolic in $G\ell_2(\mathbb{Q}_p)$ has the form $B_p = \{\begin{pmatrix}\alpha & x\\ 0 & \beta\end{pmatrix} | \alpha,\beta \in \mathbb{Q}_p^x, x \in \mathbb{Q}_p\}$. Also the admissible, irreducible representations of B_p which are trivial on the unipotent subgroup $\{\begin{pmatrix}1 & x\\ 0 & 1\end{pmatrix} | x \in \mathbb{Q}_p\}$ are of the form:

$$(3.5) \qquad \begin{pmatrix}\alpha & x\\ 0 & \beta\end{pmatrix} \xrightarrow{\mu} \mu_1(\alpha)\mu_2(\beta),$$

where μ_1, μ_2 are arbitrary quasicharacters on $\mathbb{Q}_p^x$.

Then with the data above, we see that $P \times B_p$ is a minimal parabolic subgroup of $\widetilde{S}_p \times G\ell_2(\mathbb{Q}_p)$ and that $\chi \otimes \mu$ defines an arbitrary, irreducible, admissible representation of $P \times B_p$ which is trivial on $U \times \{\begin{pmatrix}1 & x\\ 0 & 1\end{pmatrix} | x \in \mathbb{Q}_p\}$. Then we form the induced representation $\mathrm{ind}\,(\chi \otimes \mu)$, which consists of all functions $\varphi : \widetilde{S}_p \times G\ell_2(\mathbb{Q}_p) \to V_{\chi\otimes\mu}$ which are locally constant and satisfy $\varphi(g \cdot v) = (\chi \otimes \mu)(v)\varphi(g)$ for all $g \in \widetilde{S}_p \times G\ell_2(\mathbb{Q}_p)$ and $v \in P \times B_p$. (Here $V_{\chi\otimes\mu}$ is the underlying vector space associated to the representation $\chi \otimes \mu$.)

Then the main problem is to determine, for each $\chi \otimes \mu$, the space $\mathrm{Hom}_{\widetilde{S}_p \times G\ell_2(\mathbb{Q}_p)}(\pi_p, \mathrm{ind}(\chi \otimes \mu))$, that is the linear space of all intertwining operators from π_p to $\mathrm{ind}(\chi \otimes \mu)$.

The method for determining $\mathrm{Hom}_{\widetilde{S}_p \times G\ell_2}(\pi_p, \mathrm{ind}(\chi \otimes \mu))$ will now be sketched. The actual proof is fairly technical, and the details are given in [10]. We shall give here only an outline of the ideas used in [10].

Since the representation π_p of $\widetilde{S}_p \times G\ell_2(\mathbb{Q}_p)$ on $S(\mathbb{Q}_p^{2m} \times \mathbb{Q}_p^x)$ is smooth, the problem of determining $\mathrm{Hom}_{\widetilde{S}_p \times G\ell_2}(\pi_p, \mathrm{ind}(\chi \otimes \mu))$ is equivalent by Frobenius duality to finding all distributions $T : S(\mathbb{Q}_p^{2m} \times \mathbb{Q}_p^x) \to V_{\chi\otimes\mu}$ which satisfy $\langle T, \pi_p(v)\varphi\rangle = \chi \otimes \mu(v)(\langle T, \varphi\rangle)$ for all $v \in P \times B_p$ and $\varphi \in S(\mathbb{Q}_p^{2m} \times \mathbb{Q}_p^x)$

The first step is then to find all such T which satisfy $\langle T,\pi_p(v)\varphi\rangle = \langle T,\varphi\rangle$ for all $v \in U \times \{\begin{pmatrix}1 & x\\ 0 & 1\end{pmatrix} | x \in \mathbb{Q}_p\}$ (i.e. the $U \times \{\begin{pmatrix}1 & x\\ 0 & 1\end{pmatrix} | x \in \mathbb{Q}_p\}$ invariant distributions). We note here that the space of such distributions is the dual space of the Jacquet module $S(\mathbb{Q}_p^{2m} \times \mathbb{Q}_p^x)/S(\mathbb{Q}_p^{2m} \times \mathbb{Q}_p^x)_{U \times \{\begin{pmatrix}1 & x\\ 0 & 1\end{pmatrix} | x\in\mathbb{Q}_p\}}$ (see [2, p. 24] for the definition of Jacquet module, etc.).

A simple exercise shows that the space of $\{\begin{pmatrix}1 & x\\ 0 & 1\end{pmatrix} | x \in \mathbb{Q}_p\}$ invariant distributions in $\mathrm{Hom}\,(S(\mathbb{Q}_p^{2m} \times \mathbb{Q}_p^x), V_{\chi\otimes\mu})$ coincides with the space of $T \in \mathrm{Hom}(S(\mathbb{Q}_p^{2m} \times \mathbb{Q}_p^x, V_{\chi\otimes\mu})$ so that $\mathrm{supp}(T) \subseteq \Gamma_0 = \{(X,t) | \langle X,X\rangle = 0\}$. The set Γ_0 is the light cone in $\mathbb{Q}_p^{2m} \times \mathbb{Q}_p^x$. However, we note that the set $\{X \in \mathbb{Q}_p^{2m} | \langle X,X\rangle = 0\}$ is "projectively" of the form $\widetilde{S}_p/\widetilde{P}$ where $\widetilde{P}$ is a maximal parabolic in $\widetilde{S}_p$. Then to find the U invariant distributions supported in Γ_0, it simply becomes a matter of using the Bruhat decomposition in $\widetilde{S}_p/\widetilde{P}$ relative to U and then applying Bruhat's theory of intertwining distributions. Thus the problem reduces to finding certain homogeneous distributions on a fixed Bruhat cell relative to the group $Z(A) \times \{\begin{pmatrix}\alpha & 0\\ 0 & \beta\end{pmatrix} | \alpha,\beta \in \mathbb{Q}_p^x\}$. However, we note that there is a certain rigidity between the actions of $Z(A)$ and $\{\begin{pmatrix}\alpha & 0\\ 0 & \beta\end{pmatrix} | \alpha,\beta \in \mathbb{Q}_p^x\}$ on a fixed Bruhat cell! Thus, summarizing the first qualitative result, we have the following Proposition.

<u>Proposition</u> 3.1. ("<u>Multiplicity One</u>"). <u>Except for a finite number of</u> χ, <u>we have</u>

$$\dim(\mathrm{Hom}_{\widetilde{S}_p \times G\ell_2(\mathbb{Q}_p)}(\pi_p, \mathrm{ind}(\chi \otimes \mu))) \leq 1.$$

One notes that in the process of proving this Proposition, one finds that for the space above to be actually <u>nonzero</u> a very precise relationship must be satisfied between χ and μ. This is connected to the relationship between the action of $Z(A)$ and $\{\begin{pmatrix}\alpha & 0\\ 0 & \beta\end{pmatrix} | \alpha,\beta \in \mathbb{Q}_p^x\}$ on a fixed Bruhat cell in Γ_0. In any case, we are interested not in all possible intertwining operators, but in a large enough family to separate functions in the continuous spectrum of π_p. To this end, we introduce the family of intertwining maps

(3.6) $$Z(\pi_p(g_1,g_2)^{-1}\varphi,\lambda_1,\lambda_2) = \frac{1}{\operatorname{denom}\rho(\lambda_1)}\int_{\mathbb{Q}_p\times\mathbb{Q}_p^x}\pi_p(g_1,g_2)^{-1}(\varphi)(tv_1,s)\lambda_1(t)\lambda_2(s)\,d^x(t)\,d^x(s),$$

where v_1 is a nonzero $\langle\ ,\ \rangle$ isotropic vector in $\mathbb{Q}_p^{2m}$, λ_1 and λ_2 are quasicharacters on $\mathbb{Q}_p^x$ of the form $\lambda_1(x) = \alpha_1(x)|x|_p^{s_1}$ and $\lambda_2(x) = \alpha_2(x)|x|_p^{s_2}$, with α_1, α_2 unitary characters on $\mathbb{Q}_p^x$, and s_1 and s_2, complex numbers with $\operatorname{Re}(s_1) > 0$, and $d^x(t)$ and $d^x(s)$ are respectively the measures $\frac{dt}{|t|_p}$ and $\frac{ds}{|s|_p}$ with dt, ds additive Haar measures on $\mathbb{Q}_p$. Moreover $\rho(\lambda_1) = \rho(\alpha_1,s_1)$ is the usual Tate "gamma" factor (see [10]). The distribution defined by Z above is essentially supported on the smallest Bruhat cell in Γ_0, i.e. the variety $Q_p \cdot v_1 \times \mathbb{Q}_p^x$. But following the usual methods, we deduce that, as a function of (s_1, s_2), $Z(\ldots,\lambda_1,\lambda_2)$ admits an analytic continuation to all of $\mathbb{C}\times\mathbb{C}$. We denote the continuation of Z also by Z. Then the intertwining operator defined by Z gives a nonzero map from π_p to $\operatorname{Ind}(\chi\otimes\mu)$, where χ is specified by $\chi_1 = \lambda_1^{-1}$, $\chi_2 = \ldots = \chi_\nu = 1$, $\chi_{\nu+1} = \lambda_2|\ |_p^{(m-1)/2}$, and $\sigma = 1$, and μ is specified by $\mu_1 = \lambda_1\lambda_2^{-1}|\ |_p^{-m+1/2}\langle\ |\Delta_Q\rangle$ and $\mu_2 = \lambda_2^{-1}|\ |_p^{1/2}$ (where $\Delta_Q = (-1)^m\det(S_L)_p$ is the local discriminant of $(S_L)_p$ and $\langle\ |\ \rangle$ is the Hilbert symbol of $\mathbb{Q}_p^x$).

We then let $T = \bigcap_{\lambda_1,\lambda_2} \operatorname{Ker}(Z(\ ,\lambda_1,\lambda_2))$ where $\operatorname{Ker}(Z(\ ,\lambda_1,\lambda_2))$ is the kernel of the operator $Z(\ ,\lambda_1,\lambda_2)$ and the intersection varies over all pairs of quasicharacters λ_1 and λ_2. Then T is a $\tilde{S}_p \times G\ell_2(\mathbb{Q}_p)$

invariant subspace of $S(\mathbb{Q}_p^{2m} \times \mathbb{Q}_p^x)$.

Then we can prove that

<u>Proposition</u> 3.2. ([10])

(1) <u>Let</u> χ <u>and</u> μ <u>be related as in the above paragraph</u>. <u>Aside from a finite number of</u> χ, <u>we have that</u> $Z(\ ,\lambda_1,\lambda_2)$ <u>is the unique</u> (<u>up to scalars</u>) <u>element belonging to</u> $\mathrm{Hom}_{\tilde{S}_p \times G\ell_2(\mathbb{Q}_p)}(\pi_p, \mathrm{ind}(\chi \otimes \mu))$.

(2) <u>The representation of</u> π_p <u>restricted to</u> T <u>is quasicuspidal</u>. <u>That is</u>, $(T/T_W) = \{0\}$ <u>for any</u> W, <u>the unipotent radical of a parabolic subgroup in</u> $\tilde{S}_p \times G\ell_2(\mathbb{Q}_p)$.

The main idea in proving (2) of the above Proposition is that if $\varphi \in T$, then φ vanishes in a neighborhood of each point $(X,t) \in \Gamma_0$; hence φ is perpendicular to all distributions supported on Γ_0. This implies that $\varphi \in S(\mathbb{Q}_p^{2m} \times \mathbb{Q}_p^x)_{\{(\begin{smallmatrix}1 & x\\ 0 & 1\end{smallmatrix})|x \in \mathbb{Q}_p\}} \cap T = T_{\{(\begin{smallmatrix}1 & x\\ 0 & 1\end{smallmatrix})|x \in \mathbb{Q}_p\}} \subseteq T_W$ for all unipotent radicals W in $\tilde{S}_p \times G\ell_2(\mathbb{Q}_p)$, where the projection of W on the $G\ell_2(\mathbb{Q}_p)$ factor is $\{(\begin{smallmatrix}1 & x\\ 0 & 1\end{smallmatrix})|x \in \mathbb{Q}_p\}$.

Finally, we are now in a position to determine the generators of the ideal I_p. Recalling the form of $(S_L)_p$ given above, we deduce easily that the space $S(\mathbb{Q}_p^{2m} \times \mathbb{Q}_p)^{U'_p \times G\ell_2(\mathcal{O}_p)}$ (the set of $U'_p \times G\ell_2(\mathcal{O}_p)$ invariant vectors in $S(\mathbb{Q}_p^{2m} \times \mathbb{Q}_p^x)$) is nonzero. Then we have, using (2) of the above Proposition, that $T \cap S(\mathbb{Q}_p^{2m} \times \mathbb{Q}_p^x)^{U'_p \times G\ell_2(\mathcal{O}_p)} = \{0\}$. Moreover, we note that for each $\varphi \in S(\mathbb{Q}_p^{2m} \times \mathbb{Q}_p^x)^{U'_p \times G\ell_2(\mathcal{O}_p)}$, $Z(\varphi,\lambda_1,\lambda_2) \equiv 0$ for all λ_1,λ_2 which do not determine a <u>class-one</u> representation $\mathrm{ind}(\chi \otimes \mu)$ of $\tilde{S}_p \times G\ell_2(\mathbb{Q}_p)$ ($\mathrm{ind}(\chi \otimes \mu)$ is class-one if $\mathrm{ind}(\chi \otimes \mu)$ admits a nonzero $U'_p \times G\ell_2(\mathcal{O}_p)$ fixed vector). We recall the conditions imposed on χ and μ given above; then it is easy to see that $\mathrm{ind}(\chi \otimes \mu)$ is a class-one representation if and only if α_1 and α_2

are unramified unitary characters on $\mathbb{Q}_p^\times$. Thus for a given φ which is $U'_p \times G\ell_2(\mathcal{O}_p)$ invariant, we see that if $Z(\pi_p(g_1,g_2)^{-1}\varphi,\lambda_1,\lambda_2) \equiv 0$ for all $\lambda_1 = |\ |_p^{s_1}$, $\lambda_2 = |\ |_p^{s_2}$ (where s_1, s_2 range over all complex numbers) and all $(g_1,g_2) \in \tilde{S}_p \times G\ell_2(\mathbb{Q}_p)$, then $\varphi = 0$.

Suppose now that φ is $U'_p \times G\ell_2(\mathcal{O}_p)$ invariant, $f_1 \in \mathcal{H}(\tilde{S}_p//U'_p)$, and $f_2 \in \mathcal{H}(G\ell_2(\mathbb{Q}_p)//G\ell_2(\mathcal{O}_p))$, then $Z((\pi_p(f_1) - \pi_p(f_2))(\varphi),\lambda_1,\lambda_2) \equiv \{\mathrm{Trace}(\mathrm{ind}_\chi(f_1)) - \mathrm{Trace}(\mathrm{ind}_\mu(f_2))\}Z(\varphi,\lambda_1,\lambda_2)$, where $\mathrm{Trace}(\mathrm{ind}_\chi(f_1))$ and $\mathrm{Trace}(\mathrm{ind}_\mu(f_2))$ are the respective traces of the operators $\mathrm{ind}_\chi(f_1)$ and $\mathrm{ind}_\mu(f_2)$ (f_1 and f_2 are operating by convolution on the spaces ind_χ and ind_μ, respectively). Thus, to define a map $\mathrm{Cor}_p : \mathcal{H}(\tilde{S}_p//U'_p) \to \mathcal{H}(G\ell_2(\mathbb{Q}_p)//G\ell_2(\mathcal{O}_p))$, we must find, for each $f_1 \in \mathcal{H}(\tilde{S}_p//U'_p)$, a unique $\mathrm{Cor}_p(f_1) \in \mathcal{H}(G\ell_2(\mathbb{Q}_p)//G\ell_2(\mathcal{O}_p))$ so that $\mathrm{Trace}(\mathrm{ind}_\chi(f_1)) = \mathrm{Trace}(\mathrm{ind}_\mu(\mathrm{Cor}_p(f_1)))$ for all pairs $\lambda_1 = |\ |_p^{s_1}$, $\lambda_2 = |\ |_p^{s_2}$ (with s_1, s_2 ranging over all complex numbers.) We show in [10] by direct computation that this is possible (here we recall the hypotheses of Theorem 2.1 and Remark 2.2). We note that the uniqueness part of the statement is very easy to show. Indeed, suppose $\mathrm{Trace}(\mathrm{ind}_\mu(g_1)) = \mathrm{Trace}(\mathrm{ind}_\mu(g_2))$ for all μ; as s_1 and s_2 run over all complex numbers, we then obtain all the characters of the algebra $\mathcal{H}(G\ell_2(\mathbb{Q}_p)//G\ell_2(\mathcal{O}_p))$ by $\varphi \to \mathrm{Trace}(\mathrm{ind}_\mu(\varphi))$. Hence $g_1 \equiv g_2$ above.

Finally, it is clear that the generators of the ideal I_p can be constructed from the comments above as follows. Let ξ_i be a set of generators of $\mathcal{H}(\tilde{S}_p//U'_p)$. Then the set $\xi_i - \mathrm{Cor}_p(\xi_i)$ form a set of generators of I_p.

We give the Cor_p homomorphism explicitly in §4. We note here that the arguments given above work in the odd dimensional case also (with the appropriate modifications, see [10]).

§4. The Cor_p Homomorphism

To give an explicit description of the Cor_p map, we must first study the spectral theory of the Hecke algebra $\mathcal{H}(\tilde{S}_p//U'_p)$ (when $p \nmid 2q(L)$). Following [11], it is possible to describe explicitly the algebra structure of $\mathcal{H}(\tilde{S}_p//U'_p)$.

Following the notation of §3, we recall here that $\nu = m-1$ if $\dim E > 0$ and $\nu = m$ if $\dim E = 0$.

The double coset decomposition of $\tilde{S}_p$ relative to U'_p has a very simple parametrization. That is, every double coset can be put in the form $U'_p \pi^t U'_p$ where

$$(4.1) \qquad \pi^t = \begin{cases} \mathrm{diag}(p^{t_1}, p^{t_2}, \ldots, p^{t_m}, p^{t_0-t_1}, \ldots, p^{t_0-t_m}) & \text{if } \dim E = 0 \\ \mathrm{diag}(p^{t_1}, p^{t_2}, \ldots, p^{t_{m-1}}, \omega^{t_0}, \nu_p(\omega)^{t_0} p^{-t_1}, \ldots, \\ \qquad \nu_p(\omega)^{t_0} p^{-t_{m-1}}) & \text{if } \dim E > 0 \end{cases}$$

We note that, with the notation given in §3, $\mathrm{diag}(\ldots)$ is taken relative to the basis $(w_1, \ldots, w_m, \tilde{w}_1, \ldots, \tilde{w}_m)$ $((w_1, \ldots, w_{m-1}, E, \tilde{w}_1, \ldots, \tilde{w}_{m-1})$ resp.) if $\dim E = 0$ (if $\dim E > 0$). Also ω is an element in G_0 which satisfies $\mathrm{ord}_p \nu_p(\omega) = 2$, and $t = (t_1, \ldots, t_\nu, t_0)$ is a tuple of $\nu + 1$ integers, with $t_1 \geq t_2 \geq \ldots \geq t_{m-1} \geq t_0$ if $\dim E > 0$. and with $t_1 \geq t_2 \geq \ldots \geq t_{m-1} \geq \max(t_m, t_0 - t_m)$ if $\dim E = 0$. We note here that if $\dim E > 0$, then $(t_1 \ldots, t_{m-1})$ (if $\dim E = 0$, then $(t_1 \ldots, t_{m-1}, \max(t_m, t_0 - t_m))$, resp.) is the set of elementary divisors of $\pi^t((L_0)_p)$ relative to $(L_0)_p$ and that

$$n(\pi^t(L_0)_p) = \begin{cases} p^{t_0} & \text{if } \dim E = 0 \\ p^{2t_0} & \text{if } \dim E > 0 \end{cases}, \quad \text{when } \pi^t(L_0)_p \subseteq (L_0)_p.$$

We let $\mathfrak{t}^{(i)}$ be the characteristic function of the coset $U_p \pi^{(i)} U'_p$, where $(i) = (\overbrace{1,1,\ldots,}^{i}0,\ldots,0)$, $1 \le i \le \nu$, and $\phi^{(0)'}$, the characteristic function of $U'_p \pi^{(0)'} U'_p$ where $(0)' = (1,\ldots,1)$. Then if $\dim E > 0$, we have that $\mathcal{H}(\tilde{S}_p//U'_p)$ is generated by $\phi^{(i)}$, $1 \le i \le m-1$, and $\phi^{(0)'}$; that is, $\mathcal{H}(\tilde{S}_p//U'_p)$ is isomorphic as an algebra to the ring of polynomials $\mathbb{C}[\phi^{(1)},\ldots,\phi^{(m-1)},\phi^{(0)'},(\phi^{(0)'})^{-1}]$. On the other hand, if $\dim E = 0$, then we let $\phi^{(m)'}$ and $\phi^{(m)''}$ be the characteristic functions of $U'_p \pi^{(m)'} U'_p$ and $U'_p \pi^{(m)''} U'_p$, respectively, where $m' = (\overbrace{1,\ldots,1}^{m-1},0,1)$ and $(m)'' = (1,1,\ldots,1,2)$. Then (if $\dim E = 0$) $\mathcal{H}(\tilde{S}_p//U'_p)$ is isomorphic as an algebra to the ring of polynomials $\mathbb{C}[\phi^{(1)},\ldots,\phi^{(m-2)},\phi^{(0)'},\phi^{(m)'},\phi^{(m)''},(\phi^{(m)''})^{-1}]$.

For any two π^{t_1} and π^{t_2}, we let $c(t_1,t_2)$ be the cardinality of the set $U'_p\backslash U'_p \pi^{t_2} U \cap U'_p \pi^{t_1} U'_p$, where U is the unipotent radical of the minimal parabolic $\tilde{P}$ (defined in §3) of $\tilde{S}_p$.

We also let δ be the Jacobian of the action of A on U. That is, $\delta(a) = \det(Ad(a)|_{\underset{\sim}{u}})$ where Ad is the adjoint action of A on the Lie algebra $\underset{\sim}{u}$ of U.

We then recall that every homomorphism of $\mathcal{H}(\tilde{S}_p//U'_p)$ is given by the map $\varphi \rightsquigarrow \mathrm{Trace}(\mathrm{ind}_\chi(\varphi))$, where χ is an unramified character of $Z(A)$ (using 3.4) of the form

$$(4.2) \qquad (\xi_1,\ldots,\xi_\nu,g) \to \Big(\prod_{i=1}^{i=\nu} |\xi_i|_p^{s_i+i-m}\Big)|\nu_p(g)|^{\Delta(s_0)}$$

where $\Delta(s_0) = (m/4)(\dim E + (m-1)) + s_0$ and $s = (s_1,\ldots,s_\nu,s_0)$ is a tuple of $(\nu+1)$ complex numbers.

On the other hand by the well known theory of spherical functions, one knows that there exists a unique function φ_s on $\tilde{S}_p$ which satisfies $\varphi_s(k_1 g k_2) = \varphi(g)$ for all $k_1,k_2 \in U'_p$, $g \in \tilde{S}_p$, and so that $\varphi_s(e) = 1$ and

$$(4.3) \qquad \mathrm{Trace}(\mathrm{ind}_\chi(\varphi)) = \int_{\tilde{S}_p} \varphi(x)\varphi_s(x)dx \ .$$

We note that $\mathrm{Trace}(\mathrm{ind}_\chi(f) = \hat{f}(-s)$, where $\hat{f}$ is the Satake Fourier transform of f given in [11]. We then define the following set of trigonometric polynomials, i.e. $X_t(s_1,\ldots,s_\nu,s_0) = \sum_{i_1<i_2<\ldots<i_t} p^{\pm s_1 \pm s_2 \pm s_3 \pm \cdots \pm s_{i_t}}$

where $1 \leq t \leq \nu$,

$$X_0(s_1,\ldots,s_\nu,s_0) = p^{-2s_0}p^{-(s_1+s_2+\ldots+s_{i_\nu})},$$

$$X'_\nu(s_1,\ldots,s_\nu,s_0) = p^{-s_0} \sum_{\substack{i_1<i_2<\ldots<i_t \\ \text{with } t=\nu-2j}} p^{-(s_{i_1}+s_{i_2}+\ldots+s_{i_t})},$$

$$X''_\nu(s_1,\ldots,s_\nu,s_0) = p^{-s_0} \sum_{\substack{i_1<i_2<\ldots<i_t \\ \text{with } t=\nu-(2j+1)}} p^{-(s_{i_1}+s_{i_2}+\ldots+s_{i_t})}.$$

Thus using the formalism of [11], (if $\dim E > 0$) we see that $\hat{\mathfrak{m}}^{(i)}(s) = \lambda_{(i)(0)} + \lambda_{(i)(1)}X_1 + \ldots + \lambda_{(i)(i)}X_i$, where $\lambda_{(i)(j)} = c((i),(j))\delta^{-1/2}\mathfrak{m}^{(j}$ Also we have that $\hat{\mathfrak{m}}^{(0)}(s) = X_0$. On the other hand, if $\dim E = 0$, then $\hat{\mathfrak{m}}^{(i)}$ is the same as above for $1 \leq i \leq m-2$. Moreover, $\hat{\mathfrak{m}}^{(0)'}(s) = \lambda_{(0)'(m)'}X''_m + \lambda_{(0)'(0)'}X'_m$, $\hat{\mathfrak{m}}^{(m)'}(s) = \lambda_{(m)'(m)'}X''_m$, and $\hat{\mathfrak{m}}^{(m)''}(s) = \lambda_{(m)''(m)''}X_0$.

Then we set $X_i(s_1,\ldots,s_\nu) = (p^{s_1}+p^{-s_1})X_{i-1}(s_2,\ldots,s_\nu) + R_i(s_2,\ldots,s_\nu)$, where R_i does not involve terms of the form p^{X_i} or p^{-X_i} (note, if $m = 1$, then $R_1 = 0$ and $P_0 = 1$). Then we also define $A^w_{\text{even}}(s_1,\ldots,s_{m-1}) =$

$$\sum_{\substack{i_1,i_2<\ldots<i_t \\ \text{with } t=w-2j \\ j\geq 0}} p^{(s_{i_1}+s_{i_2}+\ldots+s_{i_t})}.$$

Moreover we let $A_i = \sum_{j=1}^{j=i} \lambda_{(i)(j)} P_{j-1}(m-2,\ldots,m-\nu)$ and

$B_i = \lambda_{(i)(0)} + \sum_{j=1}^{j=i} \lambda_{(i)(j)} R_j(m-2,\ldots,m-\nu)$.

We also recall that the algebra $\mathcal{H}(G\ell_2(\mathbb{Q}_p)//G\ell_2(\mathcal{O}_p))$ is generated by the characteristic functions X_p and Z_p of the double cosets $G\ell_2(\mathcal{O}_p)\begin{pmatrix} p & 0 \\ 0 & 1\end{pmatrix}G\ell_2(\mathcal{O}_p)$ and $G\ell_2(\mathcal{O}_p)\begin{pmatrix} p & 0 \\ 0 & p\end{pmatrix}G\ell_2(\mathcal{O}_p)$, i.e. $\mathcal{H}(G\ell_2(\mathbb{Q}_p)//G\ell_2(\mathcal{O}_p))$ is isomorphic as an algebra to $\mathbb{C}[X_p, Z_p, Z_p^{-1}]$. We let Y_p be the characteristic function of $G\ell_2(\mathcal{O}_p)\left(\begin{smallmatrix} p & 0 \\ 0 & p\end{smallmatrix}-1\right)G\ell_2(\mathcal{O}_p)$. Then an easy exercise shows that $Y_p = \frac{1}{p}(X_p * X_p * Z_p^{-1}) + (p-3)$. Also we let W_p be the characteristic function of the double coset $G\ell_2(\mathcal{O}_p)\left(\begin{smallmatrix} 1 & 0 \\ 0 & p\end{smallmatrix}-1\right)G\ell_2(\mathcal{O}_p)$. Then we have that $W_p = X_p * Z_p^{-1}$.

Finally with all this notation set forth, we have from [10] that the Cor_p homomorphism is given as follows.

(A) Let $\dim E > 0$. Then

$\mathrm{Cor}_p(\Uparrow^{(0)'}) = \lambda_p Z_p^{-1}$ and

$\mathrm{Cor}_p(\Uparrow^{(i)}) = \lambda_p A_i Y_p + B_i - (p-1)\lambda_p A_i$ for $i = 1,\ldots,m-1$,

where $\lambda_p = \left(\frac{(-1)^m \det S_L}{p}\right)$

(B) Let $\dim E = 0$. Then with $\left(\frac{(-1)^m}{p}\right) = 1$,

$\mathrm{Cor}_p(\Uparrow^{(m)''}) = Z_p^{-1}$

$\mathrm{Cor}_p(\Uparrow^{(i)}) = A_i Y_p + B_i - (p-1)A_i$ for $i = 1,\ldots,m-2$

$\mathrm{Cor}_p(\Uparrow^{(m)'}) = \lambda_{(m)'(m)'} C_m p^{(m/2)-1-(m/4)(m-1)} W_p$

$\mathrm{Cor}_p(\Uparrow^{(0)'}) = [\lambda_{(0)'(m)'} + \lambda_{(0)'(0)'}] C_m p^{(m/2)-1-(m/4)(m-1)} W_p$

where

$C_m = A^{m-1}_{\mathrm{even}}(m-2,m-3,\ldots,0)$.

Remark 4.1. We note that if $(\frac{(-1)^m}{p}) = -1$ (with $\dim E = 0$ and $p \neq 2$), then the Cor_p homomorphism cannot be defined on all of $\mathcal{H}(\tilde{S}_p//U'_p)$. However it is possible to define Cor_p on $\mathcal{H}'(\tilde{S}_p//U'_p) = \{\varphi \in \mathcal{H}(\tilde{S}_p//U'_p) | \operatorname{supp}(\varphi) \subseteq \{x \in \tilde{S}_p | \nu_p(x) \in (k^x)^2\}\}$. Indeed, it is straightforward to show that $\mathcal{H}'$ is generated by $\pi^{(i)}$, $i = 1,\ldots,m-2$, $(\pi^{(0)'})^2$, $(\pi^{(m)'})^2$, $\pi^{(0)'} \cdot \pi^{(m)'}$, $\pi^{(m)''}$, and $(\pi^{(m)''})^{-1}$. Then the Cor_p homomorphism defined on $\mathcal{H}'$ is given as follows:

$$(C)\quad \begin{aligned} &Cor_p(\pi^{(m)''}) = -Z_p^{-1} \\ &Cor_p(\pi^{(i)'}) = -A_i Y_p + B_i + (p-1)A_i \quad \text{for } i = 1,\ldots,m-2 \\ &Cor_p((\pi^{(0)'})^2) = \alpha_p[\tfrac{1}{p} W_p^2 - 4Z_p^{-1}] \\ &Cor_p((\pi^{(m)'})^2) = \beta_p[\tfrac{1}{p} W_p^2 - 4Z_p^{-1}] \\ &Cor_p(\pi^{(0)'}\pi^{(m)'}) = \gamma_p[\tfrac{1}{p} W_p^2 - 4Z_p^{-1}] \end{aligned}$$

where $\alpha_p = (\lambda_{(0)'(m)'} + \lambda_{(0)'(0)'})^2 C_m^2 p^{m-(m/2)(m-1)-1}$, $\beta_p = \lambda^2_{(m)'(m)'} C_m^2 p^{m-(m/2)(m-1)}$ and $\gamma_p = \lambda_{(m)'(m)'}(\lambda_{(0)'(m)'} + \lambda_{(0)'(0)'})C_m^2 p^{m-(m/2)(m-1)-1}$.

§5. Applications and Open Problems

(I) One of the easiest applications of the formulae given in §4 is to derive Siegel's formula similar to the formula in Remark 1.1. Indeed, we assume that L is unimodular (so that $q(L) = 1$ and $\det S_L = 1$). Hence $m = 4k$. Let $L_1,\ldots,L_h$ be representatives of the $G\ell_{2m}(\mathbb{Z})$ classes in the S genus of L. In Theorem 2.1, we let $f = 1$, the constant function 1 in $F(U'_{L_0}\backslash \tilde{S}^1_A/\tilde{S}_Q)$. Then we have that $i_p(\varphi) * f = c(\varphi)f$ for all $\varphi \in \mathcal{H}(\tilde{S}_p//U'_p)$. In particular, if φ is the characteristic function of a double coset $U'_p\alpha_p U'_p$, then $c(\varphi) = \text{degree } [\alpha_p] =$ the number of left (right) U'_p cosets in $U'_p\alpha U'_p$. On the other hand, one has

$\mathfrak{L}(1) = \sum_{i=1}^{h} \frac{1}{\varepsilon_{L_i}} \theta_{L_i}$. Let p be a prime so that $p \neq 2$ and $(S_L)_p$ is a totally split form at $\mathbb{Q}_p$ (e. g. when $p \equiv 1 \bmod 4$). But then, using Theorem 2.1, we have that $\mathfrak{L}(\mathbb{H}^{(m)'} * 1) = \mathrm{Cor}(\check{\mathbb{H}}^{(m)'}) * \mathfrak{L}(1) = p^{(m/2)-1-(m/8)(m-1)} \lambda_{(m)'(m)'} C_m X_p * \mathfrak{L}(1)$. Thus we deduce that

$$(5.1) \qquad T(p)\Big(\sum_{i=1}^{h} \frac{1}{\varepsilon_{L_i}} \theta_{L_i}\Big) = \frac{\mathrm{degree}(\mathbb{H}^{(m)'})}{p^{-m(m-1)/4}\, \lambda_{(m)'(m)'} C_m}\Big(\sum_{i=1}^{h} \frac{1}{\varepsilon_{L_i}} \theta_{L_i}\Big).$$

Then it becomes a matter of evaluating $\mathrm{degree}(\mathbb{H}^{(m)'})$, $\lambda_{(m)'(m)'}$, and C_m. Indeed, one knows that $\mathrm{degree}(\mathbb{H}^{(m)'}) = \frac{1}{2} \cdot$ {the number of maximal isotropic subspaces in $\mathbb{F}_p^{2m}$ relative to the form $\sum_{i=1}^{i=m} x_i y_{2m-i}$}. But, by well known arguments, the last number equals $2 \prod_{i=1}^{i=m-1} \left(\frac{p^{2i}-1}{p^i-1}\right)$.

On the other hand, by a simple combinatorics argument, $C_m = \prod_{i=1}^{m-2} (1+p^i)$. Also we have that $\lambda_{(m)'(m)'} = p^{m(m-1)/2}$. Thus the eigenvalue for $\mathfrak{L}(1)$ relative to $T(p)$ is $(p^{m-1}+1)$ (recall here that $p \neq 2$ and $(S_L)_p$ must be totally split at $\mathbb{Q}_p$).

Thus, in any case, we can write $\mathfrak{L}(1) = f_1+f_2+\ldots+f_j$, where each f_i is an eigenfunction for all $T(p)$, f_1 an Eisenstein series of weight m, and $f_2,\ldots,f_j$ linearly independent cusp forms. But from above, this means that the eigenvalue of f_i for each $T(p)$ ($p \neq 2$ and $(S_L)_p$ totally split at $\mathbb{Q}_p$) is $(p^{m-1}+1)$. But we know that for f_i, $i \geq 2$, the eigenvalue α_p^i of the Hecke operator $T(p)$ must satisfy $|\alpha_p^i| = O(p^{m/2})$. Thus each $f_i \equiv 0$ for $i \geq 2$. Hence

$$(5.2)\qquad \sum_{i=1}^{h} \frac{1}{\tilde{\epsilon}_{L_i}} \theta_{L_i}(z) = c \cdot E_m(z)$$

where E_m is the normalized Eisenstein series of weight m given by

$$(5.3)\qquad E_m(z) = \frac{1}{2\zeta(m)} \sum_{(c,d)\in\mathbb{Z}^2-(0,0)} \frac{1}{(cz+d)^m}$$

and $c = \sum_{i=1}^{h} \frac{1}{\tilde{\epsilon}_{L_i}}$.

Remark 5.1. We note here that in general (no restriction on L and p except that $p \neq 2$, $(S_L)_p$ is totally split at $\mathbb{Q}_p$, and $(\frac{(-1)^m}{p}) = 1$) $\sum \frac{1}{\tilde{\epsilon}_{L_i}} \theta_{L_i}$ is an eigenfunction of $T(p)$. In fact, a similar argument to the one given above shows that $\sum \frac{1}{\tilde{\epsilon}_{L_i}} \theta_{L_i}$ is perpendicular (relative to the Petersson inner produce) to all cusp forms in $[\Gamma_0(q(L)), \epsilon_L, m]$.

(II) We deduce from the explicit form of the Cor_p map that, in general, the space $\mathcal{M}(\text{gen } L, q(L), \epsilon_L)$ is a stable under the algebra of operators generated by $(T(p))^2$; moreover, in the case when $\dim E = 0$ and $(\frac{(-1)^m}{p}) = 1$ (case B in §4), $\mathcal{M}(\text{gen } L, q(L), \epsilon_L)$ is stable under $T(p)$! It would thus be of major interest to know the minimal set of $G\ell_{2m}(\mathbb{Z})$ inequivalent lattices L in { } with $q(L)$ dividing some q so that $\sum_L \mathcal{M}(\text{gen } L, q(L), \epsilon_L)$ is stable under $T(p)$. We note in this context the recent work [5] of Freitag.

(III) The Cor_p map given in §4 can be described in terms of the Langlands' principle of functoriality (see [10]). That is, it is possible to describe the correspondence or pairing between the representations of G_1 and G_2 that occur in the continuous spectrum of π_p (defined in §3) as a lifting in the sense of [3] between the associated Weil L-groups

${}^{L}G_1$ and ${}^{L}G_2$. (Here $G_1 = \{g \in \tilde{S}_p \mid |\nu_p(g)|_p = 1\}$ and $G_2 = SL_2(\mathbb{Q}_p)$.) The importance of this is that the correspondence given by the Cor_p map is predicted by the correspondence given at the $\{\infty\}$ place by the associated Weil representation.

BIBLIOGRAPHY

1. Arthur, J., "Selberg Trace Formula for Groups of F Rank One," Annals of Math., vol. 100, (1974), pp. 326-385.

2. Bernshtein, I. N. and Zelevinskii, A. V., "Representations of the Group $GL_n(F)$ where F is a Non-Archimedean Local Field," Russian Math. Survey, vol. 31, No. 3, (May-June, 1976), pp. 1-68.

3. Borel, A., "Automorphic L-functions," lectures at 1977 AMS Summer Institute, Corvallis.

4. Eichler, M., "Quadratische Formen und Orthogonale Gruppen," Grundlehren Series of Springer Verlag, Band 63, 1974 (2nd ed.).

5. Freitag, E., "Die Invarianz Gewissen von Thetareihen erzeugter Vektorräume unter Heckeoperatoren," Math. Zeitschift, vol. 156, (1977), pp. 141-155.

6. Gelbart, S., "Automorphic Forms on Adelic Groups," Annals of Math Series, vol. 83, Princeton University Press (1975).

7. Gelbart, S., "Weil's Representation and the Spectrum of the Metaplectic Group," Springer Lecture Notes, No. 530, 1976.

8. Howe, R., "Correspondence of Hecke Operators in Theory of θ-series," lecture at AMS Summer Institute, Corvallis, 1977.

9. Kneser, M., "Semi-Simple Algebraic Groups," Algebraic Number Theory (Cassels and Frohlich, editors), Thompson Book Company, (1967), pp. 250-264.

10. Rallis, S., "The Ideal of Relations Generated by the Weil Representation," preprint.

11. Satake, I., "Theory of Spherical Functions on Reductive Algebraic Groups over p-adic Fields," Publ. Math. Inst. Hautes Études Sci., No. 18, 1963.

12. Shimura, "Arithmetic Theory of Automorphic Forms," Tokyo-Princeton, Iwanami Shoten Publishers and Princeton University Press, (1971).

Department of Mathematics
Ohio State University
Columbus, Ohio 43210